Nanotechnology Science and Technology Series

Nanotechnology in the USA: Developments, Policies and Issues

Nanotechnology Science and Technology Series

National Nanotechnology Initiative: Assessment and Recommendations
Jerrod W. Kleike (Editor)
2009. ISBN 978-1-60692-727-4

Nanotechnology Research Collection - 2009/2010. DVD edition
James N. Ling (Editor)
2009. ISBN 978-1-60741-293-9

Nanotechnology Research Collection - 2009/2010. PDF edition
James N. Ling (Editor)
2009. ISBN 978-1-60741-292-2

Safe Nanotechnology
Arthur J. Cornwelle (Editor)
2009. ISBN 978-1-60692-662-8

Safe Nanotechnology in the Workplace
Nathan I. Bialor (Editor)
2009. ISBN 978-1-60692-679-6

Strategic Plan for NIOSH Nanotechnology Research and Guidance
Martin W. Lang (Author)
2009. ISBN: 978-1-60692-678-9

Nanotechnology in the USA: Developments, Policies and Issues
Carl H. Jennings (Editor)
2009. ISBN: 978-1-60692-800-4

Nanotechnology Science and Technology Series

NANOTECHNOLOGY IN THE USA: DEVELOPMENTS, POLICIES AND ISSUES

CARL H. JENNINGS
EDITOR

Nova Science Publishers, Inc.
New York

For permission to use material from this book please contact us:
Telephone 631-231-7269; Fax 631-231-8175
Web Site: http://www.novapublishers.com

LIBRARY OF CONGRESS CATALOGING-IN-PUBLICATION DATA

ISBN: 978-1-60692-800-4

Available upon request.

Published by Nova Science Publishers, Inc. ✝ New York

CONTENTS

PREFACE

Nanotechnology — a term encompassing the science, engineering, and applications of submicron materials — involves the harnessing of unique physical, chemical, and biological properties of nanoscale substances in fundamentally new and useful ways. The economic and societal promise of nanotechnology has led to substantial and sustained investments by governments and companies around the world. In 2000, the United States launched the world's first national nanotechnology program. Since then, the federal government has invested more than $8 billion in nanoscale science, engineering, and technology through the U.S. National Nanotechnology Initiative (NNI). U.S. companies and state governments have invested billions more. As a result of this focus and these investments, the United States has, in the view of many experts, emerged as a global leader in nanotechnology. However, the competition for global leadership in nanotechnology is intensifying as countries and companies around the world increase their investments. Nanotechnology's complexity and intricacies, early stage of development (with commercial pay-off possibly years away), and broad scope of potential applications engender a wide range of public policy issues.

Chapter 1- Advances in nanotechnology—the ability to see, measure, and control matter at the scale of atoms and molecules—are leading to novel applications in areas ranging from aerospace to agriculture. The broad spectrum of practical and beneficial uses of nanotechnology is built on a foundation of basic research in materials, chemistry, biology, physics, and engineering supported by the National Nanotechnology Initiative (NNI).

Chapter 2- Nanotechnology — a term encompassing the science, engineering, and applications of submicron materials — involves the harnessing of unique physical, chemical, and biological properties of nanoscale substances in fundamentally new and useful ways. The economic and societal promise of nanotechnology has led to substantial and sustained investments by governments and companies around the world. In 2000, the United States launched the world's first national nanotechnology program. Since then, the federal government has invested more than $8 billion in nanoscale science, engineering, and technology through the U.S. National Nanotechnology Initiative (NNI). U.S. companies and state governments have invested billions more. As a result of this focus and these investments, the United States has, in the view of many experts, emerged as a global leader in nanotechnology. However, the competition for global leadership in nanotechnology is intensifying as countries and companies around the world increase their investments.

Nanotechnology's complexity and intricacies, early stage of development (with commercial pay-off possibly years away), and broad scope of potential applications engender a wide range of public policy issues. Maintaining U.S. technological and commercial leadership in nanotechnology poses a variety of technical and policy challenges, including development of technologies that will enable commercial scale manufacturing of nanotechnology materials and products; environmental, health, and safety (EHS) concerns; and maintenance of public confidence in its safety.

Congress established programs, assigned responsibilities, and initiated research and development (R&D) related to these issues in the 21st Century Nanotechnology Research and Development Act of 2003 (P.L. 108-153). While many provisions of this act have no sunset provision, FY2008 is the last year of agency authorizations included in the act. Consideration may be given to reauthorization of this act in 2008.

Proponents of the NNI assert that nanotechnology is one of the most important emerging and enabling technologies and that U.S. competitiveness, technological leadership, national security, and societal interests require an aggressive approach to the development and commercialization of nanotechnology.

Critics of the NNI voice concerns that reflect disparate underlying beliefs. Some critics assert that the government is not doing enough to move technology from the laboratory into the marketplace. Others argue that the magnitude of the public investment may skew what should be market-based decisions in research, development, and commercialization. Still other critics say that the inherent risks of nanotechnology are not being addressed in a timely or effective manner.

From the NNI's inception through FY2008, Congress has appropriated a total of $8.5 billion for NNI activities. NNI funding in FY2008 is estimated to be $1.491 billion. For FY2009, President Bush requested $1.527 billion for the NNI.

Chapter 3- Nanotechnology — a term encompassing nanoscale science, engineering, and technology — is focused on understanding, controlling, and exploiting the unique properties of matter that can emerge at scales of one to 100 nanometers. A key issue before Congress regarding nanotechnology is how best to protect human health, safety, and the environment as nanoscale materials and products are researched, developed, manufactured, used, and discarded. While the rapidly emerging field of nanotechnology is believed by many to offer significant economic and societal benefits, some research results have raised concerns about the potential adverse environmental, health, and safety (EHS) implications of nanoscale materials.

Some have described nanotechnology as a two-edged sword. On the one hand, some are concerned that nanoscale particles may enter and accumulate in vital organs, such as the lungs and brains, potentially causing harm or death to humans and animals, and that the diffusion of nanoscale particles in the environment might harm ecosystems. On the other hand, some believe that nanotechnology has the potential to deliver important EHS benefits such as reducing energy consumption, pollution, and greenhouse gas emissions; remediating environmental damage; curing, managing, or preventing diseases; and offering new safety-enhancing materials that are stronger, self-repairing, and able to adapt to provide protection.

Stakeholders generally agree that concerns about potential detrimental effects of nanoscale materials and devices — both real and perceived — must be addressed to protect and improve human health, safety, and the environment; enable accurate and efficient risk assessment, risk management, and cost-benefit trade-offs; foster innovation and public confidence; and ensure that society can enjoy the widespread economic and societal benefits that nanotechnology may offer. Congressionally-mandated reviews of the National Nanotechnology Initiative (NNI) by the National Research Council and the President's

Council of Advisors on Science and Technology have concluded that additional research is required to make a rigorous risk assessment of nanoscale materials.

Chapter 4- The National Nanotechnology Initiative (NNI), administered by the Office of Science and Technology Policy (OSTP), is a multiagency effort intended to coordinate the nanotechnologyrelated activities of 25 federal agencies that fund nanoscale research or have a stake in the results. Nanotechnology is the ability to control matter at the scale of a nanometer—one billionth of a meter. A key research area funded by some federal agencies relates to potential environmental, health, and safety (EHS) risks that may result from exposure to nanoscale materials. Because of concerns about federal efforts to fund and prioritize EHS research, GAO was asked to determine (1) the extent to which selected agencies conducted such research in fiscal year 2006; (2) the reasonableness of the agencies' and the NNI's processes to identify and prioritize such federal research; and (3) the effectiveness of the agencies' and the NNI's process to coordinate this research. GAO reviewed quantitative and qualitative data from five federal agencies that provided 96 percent of fiscal year 2006 funding for EHS research.

Chapter 5- In March 2008, GAO issued a report entitled Nanotechnology: Better Guidance Is Needed to Ensure Accurate Reporting of Federal Research Focused on Environmental, Health, and Safety Risks (GAO-08-402). In this report, GAO reviewed the National Nanotechnology Initiative (NNI), a multiagency effort administered by the Office of Science and Technology Policy (OSTP). The NNI coordinates the nanotechnology-related activities of 25 federal agencies that fund nanoscale research or have a stake in the results. A key research area funded by some agencies related to studying the potential environmental, health, and safety (EHS) risks that may result from exposure to nanoscale materials. For this testimony statement, GAO was asked to summarize the findings of its March 2008 report, focusing on (1) the extent to which selected agencies conducted EHS research in fiscal year 2006; (2) the reasonableness of the agencies' and the NNI's processes to identify and prioritize EHS research; and (3) the effectiveness of the agencies' and the NNI's process to coordinate EHS research.

Chapter 6- Scientists and engineers can now examine, design, and manipulate materials at the molecular level, termed "nanoscale," between 1 and 100 billionths of a meter. The U.S. government has invested heavily to ensure that American industry remains a global leader in the field, because the products of nanotechnology are seen to have great economic potential and offer possible solutions to national problems ranging from energy efficiency to detection of agents of biological warfare. Optimism about nanotechnology is tempered, however, by concerns about the unknown potential of nanoscale materials to harm the environment and human health. Some have called for federal regulation of potential environmental, human health, and safety (EHS) risks, arguing that the lack of federal EHS regulations increases the risks of unanticipated adverse consequences due to human or environmental exposure to engineered nanomaterials. The cost of such consequences would depend on their actual, as well as publicly perceived, severity, frequency, and reversibility. The cost to the nanotechnology industry could be great, if consumers responded to a potential threat of harm by indiscriminately rejecting all products of nanotechnology, rather than the offending nanomaterial or an individual application. Others oppose federal regulatory requirements, arguing that they might unnecessarily delay the environmental, health, and economic rewards expected from nanotechnology.

Questions about the need for, and ideal form of, nanotechnology regulations are exceedingly difficult to address, given the current state of scientific understanding of engineered nanoscale materials. This chapter considers certain challenges faced by scientists, entrepreneurs, and government officials involved with nanotechnology research, as they strive to define the characteristics of nanomaterials, the potential EHS risks, and how they might be addressed. Challenges include the wide variety of nanomaterials and applications; lack of basic information about their properties; lack of conventions for naming, measuring and identifying nanomaterials; the proprietary nature of some critical information; the need to prioritize federal resource needs; and a possible lack of clear statutory authority or appropriate regulatory framework to anticipate or respond to any identified risks. For more information about the national nanotechnology research agenda, appropriations, and authorizing legislation, see CRS Report RL34401, The National Nanotechnology Initiative: Overview, Reauthorization, and Appropriations Issues.

These difficulties may be surmounted over time without significant legislative action, or Congress may choose to intervene. If it does, it might choose any of several approaches. Possible approaches include increasing funding for workshops in standardization or other research relevant to identifying and possibly ameliorating any environmental or human health and safety concerns associated with nanomaterials; changing the allocation of research money among agencies or the interagency research management structure; adopting a national or international research strategy; or enacting legislation that authorizes, mandates, or constrains agency actions to require information collection or to restrict production, sale, use, or disposal of nanomaterials. Each risk management approach has potential positive and negative consequences that Congress may want to consider.

In: Nanotechnology in the USA: Developments, Policies and Issues ISBN 978-1-60692-800-4
Editor: Carl H. Jennings

Chapter 1

THE NATIONAL NANOTECHNOLOGY INITIATIVE*

EXECUTIVE OFFICE OF THE PRESIDENT OFFICE OF SCIENCE AND TECHNOLOGY POLICY WASHINGTON, D.C. 20502

December 20, 2007

Dear Colleagues:

Advances in nanotechnology—the ability to see, measure, and control matter at the scale of atoms and molecules—are leading to novel applications in areas ranging from aerospace to agriculture. The broad spectrum of practical and beneficial uses of nanotechnology is built on a foundation of basic research in materials, chemistry, biology, physics, and engineering supported by the National Nanotechnology Initiative (NNI).

Now in its eighth year, the NNI expedites the responsible discovery and development of nanotechnology with the ultimate goal of improving the Nation's economy and the quality of life for all citizens. Today, the NNI involves 25 Federal agencies with diverse missions and expertise. The NNI agencies coordinate their nanotechnology-related activities in order to leverage their investments and facilitate progress in areas of individual and joint interest. The inherently multidisciplinary nature of nanotechnology is especially well served by such a multi- agency approach.

This NNI Strategic Plan outlines the goals and priorities of the initiative and describes approaches for achieving them. The plan supports leading edge research, sustains the extensive infrastructure of facilities, seeks to facilitate technology transfer, and addresses environmental, health, and societal concerns.

This document updates the first NNI Strategic Plan issued in 2004. It reflects progress in the science and in the management and coordination of such a broad effort that involves not only multiple agencies, but also industry, academia, and the public. In particular, the plan

* This is an edited, excerpted and augmented edition of a National Science and Technology Council Committee on Technology.

benefited from reviews of the NNI by the President's Council of Advisors on Science and Technology (PCAST) and the National Research Council. These outside reviews have provided valuable and complementary recommendations that have strengthened the initiative.

With the implementation of this plan, the United States will remain at the forefront of nanoscale science and engineering and a leader in achieving the economic benefits offered by the emerging technology.

Sincerely,

John H. Marburger, III
Director

VISION AND GOALS

The vision of the National Nanotechnology Initiative (NNI) is a future in which the ability to understand and control matter at the nanoscale leads to a revolution in technology and industry that benefits society.

The National Nanotechnology Initiative expedites the discovery, development, and deployment of nanoscale science and technology to serve the public good, through a program of coordinated research and development aligned with the missions of the participating agencies. In order to realize the NNI vision, the participating agencies are working collectively toward the following four goals:

- Goal 1: Advance a world-class nanotechnology research and development program.

The NNI ensures United States leadership in nanotechnology research and development by stimulating discovery and innovation. This program expands the boundaries of knowledge and develops technologies through a comprehensive program of research and development. The NNI agencies invest at the frontiers and intersections of many disciplines, including biology, chemistry, engineering, materials science, and physics. The interest in nanotechnology arises from its potential to significantly impact numerous fields, including aerospace, agriculture, energy, the environment, healthcare, information technology, homeland security, national defense, and transportation systems.

- Goal 2: Foster the transfer of new technologies into products for commercial and public benefit.

Nanotechnology contributes to United States competitiveness by improving existing products and processes and by creating new ones. The NNI implements strategies that maximize the economic benefits of its investments in nanotechnology, based on understanding the fundamental science and responsibly translating this knowledge into practical applications.

- Goal 3: Develop and sustain educational resources, a skilled workforce, and the supporting infrastructure and tools to advance nanotechnology.

A skilled science and engineering workforce, leading-edge instrumentation, and state-of-the-art facilities are essential to advancing nanotechnology research and development. Educational programs and resources are required to produce the next generation of nanotechnologists, that is, the researchers, inventors, engineers, and technicians who drive discovery, innovation, industry, and manufacturing.

- Goal 4: Support responsible development of nanotechnology.

The NNI aims to maximize the benefits of nanotechnology and at the same time to develop an understanding of potential risks and to develop the means to manage them. Specifically, the NNI pursues a program of research, education, and communication focused on environmental, health, safety, and broader societal dimensions of nanotechnology development.

INTRODUCTION

The ability to image, measure, model, and manipulate matter on the nanoscale is leading to new technologies that will impact virtually every sector of our economy and our daily lives. Nanoscale science, engineering, and technology are enabling

promising new materials and applications across many fields including healthcare, electronics, aeronautics, and energy. Realizing these possibilities requires continued research and accelerated innovation. The United States has been and is now the recognized leader in nanotechnology research and development (R&D), but this lead cannot be assumed to be permanent. Thus, the National Nanotechnology Initiative (NNI) is as important as ever to ensuring U.S. leadership in nanotechnology R&D.

The NNI has created a thriving nanoscale science and engineering R&D environment within the United States. As a result, scientific understanding of nanometer-scale phenomena has expanded enormously. An extensive network of R&D centers is already established. Commercialization resulting from NNIsupported research is growing. Yet exploiting the full value that nanotechnology offers depends on sustained R&D. Barriers to innovation and technology transfer need to be lowered. Researchers, educators, and technicians with new skills are required. Furthermore, nanotechnology must be developed responsibly.

Many indicators point towards the increasing importance of nanotechnology to the scientific and engineering communities. For example, several nanotechnology-specific journals have been launched by professional societies. The work published in these journals details the remarkable progress that has been made towards understanding nanoscale phenomena and controlling matter at nanoscale dimensions. Much of this research was executed at the nearly seventy research centers and user facilities established and supported through the NNI, beginning with the Nanobiotechnology Center at Cornell University in 2001 and continuing through the opening in May 2007 of the Center for Functional Nanomaterials at Brookhaven National Laboratory.

The NNI and nanotechnology itself are still young, and translating an emerging technology into economically viable products usually takes decades. Yet, already the nanotechnology innovation and development process has begun. For example, to date more than 4,800 patents have been identified under the nanotechnology classification created by the U.S. Patent and Trademark Office, many of which draw upon NNI-supported research.

About the NNI

The National Nanotechnology Initiative is the program established in fiscal year 2001 to coordinate Federal nanotechnology R&D. The NNI provides a vision of the long-term opportunities and benefits of nanotechnology. By serving as a central locus for communication, cooperation, and collaboration for all Federal agencies that wish to participate, the NNI brings together the expertise needed to guide and support the advancement of this broad and complex field. The NNI creates a framework for a comprehensive nanotechnology R&D program by establishing shared goals, priorities, and strategies, and it provides avenues for each individual agency to leverage the resources of all participating agencies.

Today the NNI consists of the individual and cooperative nanotechnology-related activities of twenty- five Federal agencies with a range of research and regulatory roles and responsibilities. Thirteen of the participating agencies have R&D budgets that relate to nanotechnology, with the reported NNI budget representing the collective sum of these. The NNI as a program does not fund research; however, it informs and influences the Federal budget and planning processes through its member agencies.

What is Nanotechnology?

Nanotechnology is the understanding and control of matter at dimensions between approximately 1 and 100 nanometers, where unique phenomena enable novel applications. Encompassing nanoscale science, engineering, and technology, nanotechnology involves imaging, measuring, modeling, and manipulating matter at this length scale.

A nanometer is one-billionth of a meter. A sheet of paper is about 100,000 nanometers thick; a single gold atom is about a third of a nanometer in diameter. Dimensions between approximately 1 and 100 nanometers are known as the nanoscale. Unusual physical, chemical, and biological properties can emerge in materials at the nanoscale. These properties may differ in important ways from the properties of bulk materials and single atoms or molecules.

About this Plan

Given the dynamic nature of the field, a periodic reexamination of the NNI Strategic Plan is essential. The 21" Century Nanotechnology Research and Development Act of 2003 calls for the NNI Strategic Plan to be updated every third year; the plan presented here updates and replaces the December 2004 plan.

A number of inputs guided this update. Independent reviews of the NNI by the President's Council of Advisors on Science and Technology and the National Research

Council of the National Academies have been strongly supportive of the NNI, its strategies, and its operations. Each review has offered specific recommendations for improving the NNI. Additional input has come from topical workshops sponsored by the NNI (see list in Appendix A). In these workshops, academic, government, and industry experts developed research recommendations for a variety of application areas, considered societal implications of nanotechnology, and addressed economic development strategies.

This plan, guided by a review of the recommendations mentioned above, represents the consensus of the participating agencies as to the goals and priorities of the NNI. The plan provides a framework within which each agency will carry out its own mission- related nanotechnology programs and that will sustain coordination of interagency activities. In addition to specifying high-level goals, the plan identifies activities aimed at accomplishing those goals. The plan also identifies major subject areas, or program component areas, in which investments are needed to ensure the success of the initiative. Finally, the plan identifies a number of representative high-impact application opportunities and critical research needs that cut across the NNI program component areas and that align with the competencies and missions of participating agencies.

This strategic plan presents an overview of the NNI and will facilitate achievement of the NNI vision by offering guidance for agency leaders, program managers, and the research community in their nanotechnology R&D investments and activities.

Program Component Areas

Program component areas (PCAs) are major subject areas under which are grouped related projects and activities. They provide an organizational framework for categorizing the activities of the NNI. Progress in these areas is critical to achieving the NNI's goals and to realizing its vision. The investment related to each PCA is reported in the annual NNI supplement to the President's Budget.

The eight program component areas of the NNI are defined as follows:

Fundamental Nanoscale Phenomena and Processes: Discovery and development of fundamental knowledge pertaining to new phenomena in the physical, biological, and engineering sciences that occur at the nanoscale. Elucidation of scientific and engineering principles related to nanoscale structures, processes, and mechanisms.

Nanomaterials: Research aimed at the discovery of novel nanoscale and nanostructured materials and at a comprehensive understanding of the properties of nanomaterials (ranging across length scales, and including interface interactions). R&D leading to the ability to design and synthesize, in a controlled manner, nanostructured materials with targeted properties.

Nanoscale Devices and Systems: R&D that applies the principles of nanoscale science and engineering to create novel, or to improve existing, devices and systems. Includes the incorporation of nanoscale or nanostructured materials to achieve improved performance or new functionality. To meet this definition, the enabling science and technology must be at the nanoscale, but the systems and devices themselves are not restricted to that size.

Instrumentation Research, Metrology, and Standards for Nanotechnology: R&D pertaining to the tools needed to advance nanotechnology research and commercialization, including next-generation instrumentation for characterization, measurement, synthesis, and design of materials, structures, devices, and systems. Also includes R&D and other activities related to development of standards, including standards for nomenclature, materials, characterization and testing, and manufacture.

Nanomanufacturing: R&D aimed at enabling scaled-up, reliable, and cost-effective manufacturing of nanoscale materials, structures, devices, and systems. Includes R&D and integration of ultra-miniaturized top-down processes and increasingly complex bottom-up or self-assembly processes.

Major Research Facilities and Instrumentation Acquisition: Establishment of user facilities, acquisition of major instrumentation, and other activities that develop, support, or enhance the Nation's scientific infrastructure for the conduct of nanoscale science, engineering, and technology R&D. Includes ongoing operation of user facilities and networks.

Environment, Health, and Safety: Research primarily directed at understanding the environmental, health, and safety impacts of nanotechnology development and corresponding risk assessment, risk management, and methods for risk mitigation.

Education and Societal Dimensions: Education-related activities such as development of materials for schools, undergraduate programs, technical training, and public communication, including outreach and engagement. Research directed at identifying and quantifying the broad implications of nanotechnology for society, including social, economic, workforce, educational, ethical, and legal implications.

The 2004 NNI Strategic Plan defined seven PCAs, including a Societal Dimensions PCA. To better understand and manage the NNI investment, the Societal Dimensions PCA defined in the 2004 plan is divided into two in this updated plan: the Environment, Health, and Safety PCA and the Education and Societal Dimensions PCA described above.

Relationship to Goals and Agencies

The PCAs represent areas of investment necessary to achieving the NNI goals. Table 1 qualitatively indicates the relative strength of the relationships between the PCAs and NNI goals. Note that the projects and activities in one or more PCAs are critical to the progress of each goal. In addition to their relationship to the NNI goals, the PCAs are strongly correlated with the missions of the agencies participating in the NNI.

The PCAs cut across the interests and activities of the participating agencies and represent areas where achieving the goals of the NNI can be expedited through interagency coordination. Table 2 shows for each participating agency which PCAs have the strongest relationships to the agency's mission, interests, and needs. For agencies with budgets for nanotechnology R&D, the strength of the relationship shown in Table 2 roughly correlates with the level of investment. For those agencies that do not have nanotechnology R&D budgets, there are nevertheless strong connections with agency missions. For example, the Food and Drug Administration (FDA) regulates a wide range of products, including foods, cosmetics, drugs, medical devices, and veterinary products, some of which utilize nanotechnology or contain nanomaterials. Thus, for the FDA to fulfill its regulatory mission, the science for understanding biological interactions of nanoscale materials must be developed and related scientific issues must be resolved.

Table 1. Relationships between the PCAs and NNI Goals

Relevance: critical / primary / secondary

PCA \ GOAL	Goal 1: Advance a world-class nanotechnology research and development program	Goal 2: Foster the transfer of new technologies into products for commercial and public benefit	Goal 3: Develop and sustain educational resources, a skilled workforce, and the supporting infrastructure and tools to advance nanotechnology	Goal 4: Support responsible development of nanotechnology
Fundamental Nanoscale Phenomena & Processes	critical	secondary	primary	secondary
Nanomaterials	critical	secondary	primary	secondary
Nanoscale Devices & Systems	primary	critical	secondary	secondary
Instrumentation Research, Metrology, & Standards	secondary	critical	secondary	primary
Nanomanufacturing	primary	critical	primary	secondary
Major Research Facilities & Instrumentation Acquisition	secondary	primary	critical	secondary
Environment, Health, & Safety	primary	secondary	primary	critical
Education & Societal Dimensions	secondary	secondary	primary	critical

Strategy and Plans: Research and Development in the National Interest

This section describes the strategy and plans for realizing the NNI vision in the context of the four NNI goals.

Goal 1: Advance a world-class research and development program

Strategic elements to achieve this goal include investing in multiple R&D pathways, mapping the leading edge of nanotechnology R&D, coordinating investments across agencies, and encouraging collaborations and other interactions.

Table 2. Relationships between the PCAs and the Missions, Interests, and Needs of NNI Agencies

■ - Primary □ - Secondary

Agency* / PCA	Fundamental Nanoscale Phenomena & Processes	Nanomaterials	Nanoscale Devices & Systems	Instrumentation Research, Metrology, & Standards	Nanomanufacturing	Major Research Facilities & Instrumentation Acquisition	Environment, Health, & Safety	Education & Societal Dimensions
BIS (DOC)	□	■	■	■	□			
CPSC	□	□	■	■	□		■	□
CSREES (USDA)	□	■	■	□	□		□	■
DOD	■	■	■	□	■	□	□	□
DOEd							□	■
DOE	■	■	□	□	□	■	□	□
DHS	■		■	■		□		
DOJ			■					
DOL		□			□		■	■
DOS	□	□	□	□	□	□	■	■
DOT	■	□	■		■		■	
DOTreas		■	■					
EPA	□	■	■	□	■		■	□
FDA (HHS)	□	□	■	□	□		■	
FS (USDA)	□	■	■	□	■		□	
IARPA	□	■	■		□			
ITC		■	■		■			
NASA	□	■	■		□	□		
NIH (HHS)	■	□	■	□			■	□
NIOSH (HHS)		□			□		■	□
NIST (DOC)	■	■	□	■	■	□	□	
NRC				■				
NSF	■	■	□	□	■	■	■	■
USGS (DOI)	■			■			■	
USPTO (DOC)		■	■	■	■			

*A glossary of abbreviations is provided in Appendix B.

R&D Investment Pathways

The NNI expands the boundaries of knowledge and develops technologies through a comprehensive program of R&D. This program invests at the frontiers and intersections of many disciplines, including biology, chemistry, engineering, materials science, and physics. Activities targeted toward this goal include support for fundamental research, use-inspired research, application research, and technology development. The research program of the NNI continues to be executed through a mix of single investigators, multi-investigator and team efforts, centers, and user facilities. Each plays an important role in the discovery and innovation process.

Single investigators. Research directed by single investigators is a key component of the NNI R&D portfolio and is considered critical to the continuing development of new concepts that can lead to breakthroughs in nanotechnology. Funding of individual investigators allows the NNI to support a broad range of ideas, including exploratory proposals that, if successful, could lead to major advances.

Multi-investigator and team efforts. Research at the frontiers and intersections of traditional disciplines enables rapid progress both in understanding nanoscale phenomena and in developing nanotechnology applications. Consequently, funding of multidisciplinary teams that bring a range of expertise to bear on a particular research topic or technology problem is a hallmark of the NNI effort. Participating agencies employ a variety of mechanisms for this type of investment.

Centers. The NNI member agencies invest substantial resources in large multidisciplinary research centers (see Figure 1, p. 12). Compared to multi-investigator research teams, these centers have a broader scope and bring together a larger group of researchers with diverse scientific expertise. Some centers have leveraged the Federal investment to obtain laboratories, instrumentation, and equipment from other sources, thus providing even more extensive opportunities and support for nanotechnology R&D.

User facilities and networks. These facilities make advanced experimental and theoretical capabilities accessible to researchers from academia, industry, and government laboratories. They bring together suites of complementary tools, including specialized, one-of-a-kind instruments, to cover a wide range of nanotechnology research, from materials synthesis through modeling. The scientific and technical staffs at these facilities are a vital component of this infrastructure, as they represent the expertise that is needed to most effectively utilize these capabilities.

Mapping the Leading Edge: Stimulating Concepts and Refining Ideas

Fostering a world-class program of research and development requires identifying challenges, recognizing opportunities, and stimulating new ideas. An effective means for mapping the leading edge of R&D is to develop a snapshot of the state of the art and use this to forecast future research directions. Towards this end, the NNI will sponsor topical and strategy-setting workshops. Such workshops bring together experts from academia, government, and industry to identify opportunities and needs, to suggest mechanisms to address these opportunities and needs, and to recommend strategies by which agency investments can be leveraged and coordinated. Current areas of common interest across several agencies that will be subjects of future workshops are:

- Sensors and nanoelectronics
- Energy
- Fate and transport of nanomaterials
- Medical and health applications

No single agency within the Federal Government has the mission or breadth of expertise to fully exploit the opportunities nanotechnology presents, to execute all the requisite research, or to provide all the necessary standards and reference data. Thus, in certain key areas, it is essential to coordinate the NNI research and development programs across multiple agencies. Often, this synergistic route is the only feasible approach. Even in areas principally of interest to a handful of agencies, benefits can be derived from formulating programs from a broader interagency perspective.

An example of this coordinated approach comes from the NNI-led study of the environmental, health, and safety implications of engineered nanoscale materials. Here, certain agencies are funding the fundamental research required to understand underlying processes, other agencies are developing instruments for field measurements, and still other agencies provide the standards necessary to operate those instruments properly. Ultimately, other policy-setting and regulatory agencies will use the information from this research to assist industry and government in developing responsible guidelines for nanotechnology research and development.

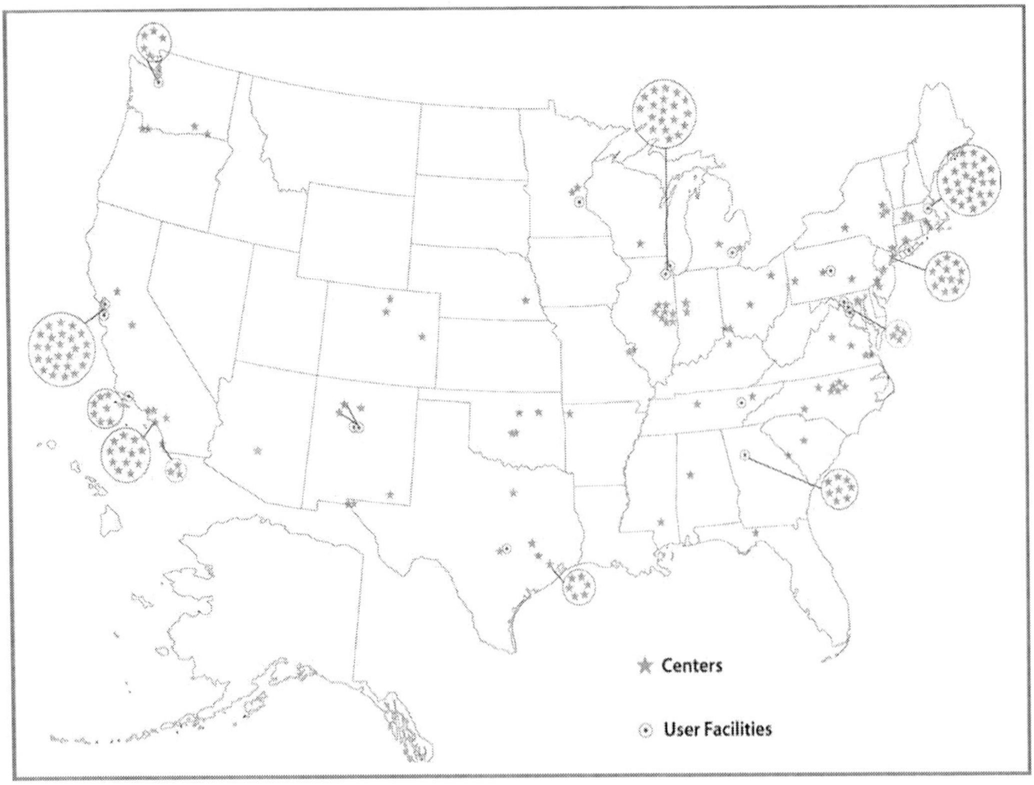

Figure 1. NNI-sponsored Centers and User Facilities (December 2007).

Successful coordination of the Federal Government's nanotechnology R&D investment requires shared priorities. In order to communicate areas of common interest across the Government, leaders from the budget, legislative affairs, and other appropriate offices of all NNI agencies will be invited to the workshops discussed in the previous section. Each workshop will include opportunities for agency representatives to discuss further means for coordinating the Federal nanotechnology R&D effort. In order to further facilitate a coordinated, interagency investment, the workshop participants will be asked to develop a list of priority research opportunities, challenges, and needs. The resulting list will be distributed to all the agencies participating in the NNI as an aid in formulating solicitations and collaborations. Additionally, members of the public will be invited to take part in these workshops as observers and to participate in associated engagement activities to discuss topics raised in the workshop that they find of interest and importance. The perspectives collected from these public engagement activities will be additional input to the NNI decision-making process.

Collaborations and Interactions

Recognizing the interdisciplinary nature of nanotechnology and the unique areas of expertise at the various Federal science agencies, the NNI agencies plan to explore opportunities for exchange of scientists across Federal agencies. Such exchanges enhance interagency connections in areas of mutual interest and expand the perspectives of Federal scientists. Scientists with this kind of experience are better equipped to assist their home agencies in furthering their research programs and provide their host agencies with fresh perspectives. The NNI will seek to facilitate targeted, cost-neutral, sabbatical-type exchanges between agencies. The possibility of similar exchanges with industry will be investigated, where appropriate. International opportunities for collaboration and other interactions will be pursued, through participation in multilateral organizations and through existing science and technology agreements with other countries. NNI member agencies also benefit from international benchmarking activities and the sharing of programmatic R&D information.

To promote and expand interagency interactions, the NNI will facilitate communications concerning ongoing and planned nanotechnology activities among the member agencies. Multiple perspectives add value to many NNI activities, such as conferences and workshops, and the participation of multiple agencies in such activities strengthens and enhances the Federal nanotechnology R&D enterprise.

Goal 2: Foster the Transfer of New Technologies into Products for Commercial and Public Benefit

In order to realize the benefits of a world-class R&D program, the NNI seeks to enhance the transfer of research results into practical applications. The technology transfer process involves partnerships between the entities that support and perform research and those with the capacity to commercialize new technologies. Advances in nanotechnology already are impacting many industry sectors.

The NNI will foster technology transfer by engaging in a coordinated way with key industry sectors and will collect and exchange information and ideas regarding each sector's technology needs. Recognizing that entrepreneurs play a large role in innovation, the NNI also seeks to provide small businesses and start-ups with access to the results of NNI-funded research. In general, the NNI will help create a business environment conducive to responsible development of nanotechnology by businesses of all sizes. Partners in this undertaking include international, regional, state, and local organizations promoting nanotechnology development, as well as professional societies, trade associations, and other nongovernmental organizations.

Coordinated Engagement with Industry

The NNI will continue to work across industry sectors to gather input and feedback on Federal research priorities and activities in order to promote commercialization for public benefit. Industry liaison groups that include government and private sector representatives promote the exchange of information on NNI research programs and industry needs. Current liaison groups with the electronics, forest products, chemical industries, and the industry research management community will continue. The formation of comparable groups with other industry sectors is under consideration.

Engagement with industry at all stages of the innovation process will aid the diffusion of nanotechnology from the laboratory to the marketplace. Interaction with industry through liaison groups and at NNI workshops helps NNI agencies to identify pre-competitive research that will be most useful in addressing industry needs.

Industry participation in the review of research proposals provides valuable input. Therefore, NNI agency proposal review panel members include representatives from industry, when appropriate and possible. In addition, participation of industry representatives, when appropriate and possible, on external review boards for NNI centers, networks, and user facilities ensures that those entities remain responsive to industry needs and promotes awareness of NNI infrastructure within the business community. Finally, industry cooperation will be essential to measuring the impact of nanotechnology in the commercial sector, for example, in collecting information regarding revenues, employment, patents, and licensing.

NNI agencies will continue to make use of existing mechanisms for technology transfer and commercial development, such as the Small Business Innovation Research (SBIR) program, the Small Business Technology Transfer (STTR) program, and the Manufacturing Technology (MANTECH) program. These programs are not specifically aimed at nanotechnology; however, agencies may highlight topics of interest to stimulate proposals in certain areas.

NNI AGENCIES COLLABORATE TO SPEED DEVELOPMENT OF CANCER THERAPIES

Nanotechnology Characterization Laboratory

A key promise of nanotechnology is for improved healthcare, including potentially revolutionary treatments for cancer. As with many nanotechnology opportunities, realizing this promise requires expertise across multiple scientific disciplines. To meet this challenge, the National Cancer Institute (NCI), working in concert with the National Institute of Standards and Technology (NIST) and the Food and Drug Administration (FDA), established the Nanotechnology Characterization Laboratory (NCL) to accelerate the development and translation of nanoscale materials and devices into cancer therapies.

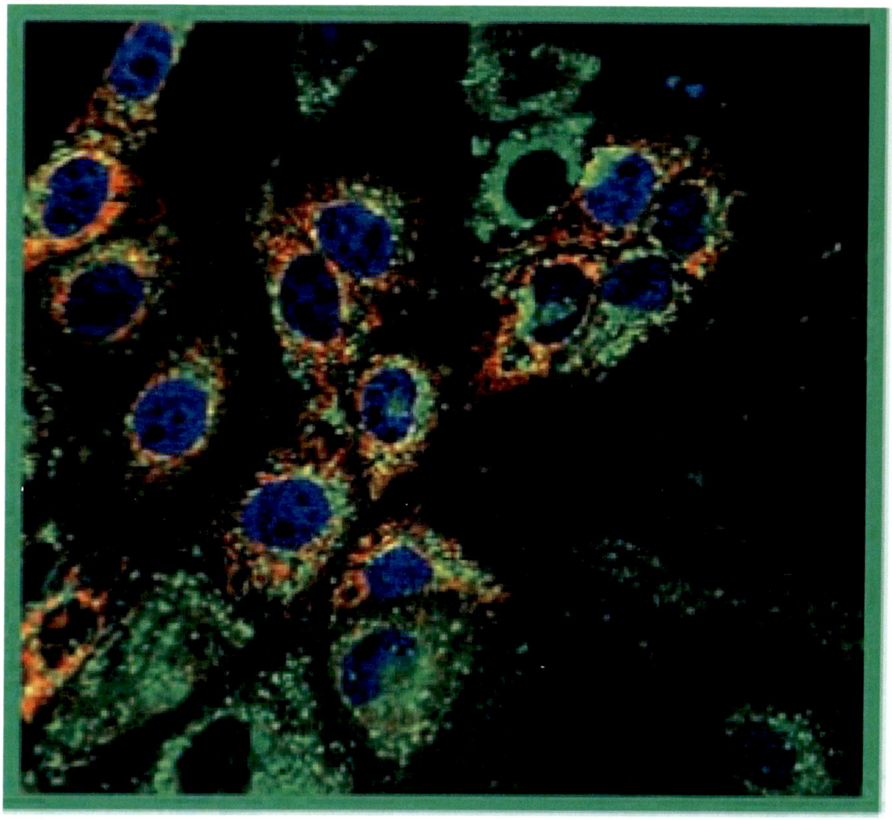

Optical confocal microscopy image of fullerenes (green) in porcine kidney cells. This method is part of an investigation of the internalization and subcellular distribution of fullerenes in cells. The nuclei of the cells have been dyed blue; the mitochondria are dyed orange in the image.(Courtesy of NCL.)

The NCL performs and standardizes preclinical characterization of nanoparticle-based platforms intended for cancer therapeutics and diagnostics. This characterization service is available to researchers from academia, government, and industry, and the data can be used in support of an Investigational New Drug filing with the FDA.The NCL also provides infrastructure support to the NCI Alliance for Nanotechnology in Cancer and serves as a national resource and knowledge base for the cancer and nanotechnology research communities.NCI, NIST, and FDA each bring unique expertise to this partnership. The NCI focuses on research directed at identifying the causes of cancer and related diseases. Its scientists investigate the many factors that contribute to human cancers; identify new targets for cancer diagnosis, treatment, and prevention; and perform state-of-the-art characterization of nanoparticle-based platforms. NIST brings state-of-the-art physical and chemical measurement instrumentation and expertise to the partnership. NIST scientists develop measurement methods, assay protocols, models, and standards for physical and chemical characterization of nanoparticles, and also apply measurement tools to characterize the physical properties of nanoparticle-based platforms submitted to the NCL. FDA reviewers are expert at evaluating safety and efficacy, and the FDA has a commitment to developing mechanisms for facilitated review of new medical technology. All three agencies support the NCL's efforts to markedly accelerate the development of nanotechnologybased products benefiting cancer patients, reduce risks associated with this development, and encourage private-sector investment in this promising area of technology. This is one example of the NNI working towards the deployment of nanotechnology for public benefit.

Creating a Favorable Business Environment

Federal agencies can make important contributions to promoting a business environment that supports the transfer of nanotechnology R&D into products that benefit the public.

Protection of intellectual property is particularly important at the early stages of technology development. The U.S. Patent and Trademark Office (USPTO) continues to improve its ability to assess inventions that are based on advances in nanotechnology. The USPTO is developing a knowledgeable corps of patent examiners and is adding detailed structure to its patent classification for nanotechnology to aid the examination process.

In addition to impacting development of nanotechnology-enabled products and processes through intellectual property protections, the Federal Government also affects related businesses through regulation. Federal agencies that have responsibilities for oversight of nanotechnology-based products are active participants in the NNI. In carrying out their responsibilities, the regulatory agencies seek to ensure may include developing guidance, as appropriate, for applying those regulations to nanotechnology-based products and disseminating that information.

As with any developing area, new information will become available over time, and additional research and new approaches may be required. Cooperative efforts among research agencies, regulators, and industry can ensure that environmental, health, and safety risks are assessed and managed, and that data gaps are addressed in a prioritized fashion that stimulates innovation, while at the same time safeguarding public welfare. The sharing of non-proprietary information on the physical and chemical properties of nanomaterials will greatly facilitate this process.

Providing channels for communication among government agencies, companies, industry groups, and the public is an additional strategic element. Such communication will provide industry with current information to support responsible development of nanotechnology and will inform the public about risks and benefits of nanotechnology for individuals, society, and the environment.

Creating an international trade environment favorable to commercialization of nanotechnology is an important thrust of the NNI. Research strategies and guidelines regarding environmental, health, and safety aspects of nanotechnology and trade-related standards development are coordinated through U.S. participation in nanotechnology activities sponsored by the Organisation for Economic Co-operation and Development, the International Organization for Standardization, and other international organizations. Supporting the development of standards for nomenclature and terminology, measurement and characterization, and guidelines for handling and processing of nanomaterials will help to create a business environment in which U.S. companies have paths to bringing their products to international as well as domestic markets.

NNI Partnerships Accelerate Discovery of a New Logic Switch
Public-Private Partnership Supporting Nanoelectronics Innovation

Experts maintain that new technology and engineering approaches are necessary for the continued vitality of the information technology industry. Nanoelectronics, the application of nanotechnology to make electronic circuits, offers revolutionary alternatives to the component technology used in existing computer circuits.

Understanding the need for transformational long-range basic research in this area, the Semiconductor Research Corporation, an industry-led research consortium, has formed the Nanoelectronics Research Initiative (NRI). The NRI has the mission of developing nanotechnology-based computing devices to succeed existing semiconductor technologies. As a result of work among experts from industry, academia, and government, the NRI has identified multiple high-priority research targets, the most pressing of which is the invention of a new logic switch by 2020.

Recognizing the importance of this industry sector, two NNI agencies, I the National Science Foundation (NSF) and the National Institute of Standards and Technology (NIST), have formed public- private partnerships with the NRI to support research in nanoelectronics with the goal of developing radically new, yet practical, successors to the components currently used in electronics and computer circuits. NSF, NIST, and the industry members of the NRI each provide funding to support research towards that goal. The research is conducted through university-based centers. Project selection is done by a panel of technical experts.

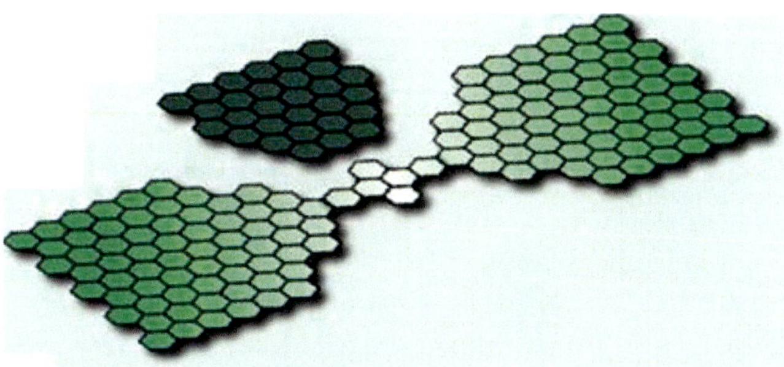

Concept of a switch made from a graphene sheet. Here a molecular quantum dot (center, white) is connected to graphene electrodes through narrow constrictions. A coplanar graphene side gate (dark green) controls charge flow through the switch. Reproduced by permission from Physics Today, © 2007 American Institute of Physics.

These government-industry-academic partnerships blend the discovery mission of NSF, the technology innovation mission of NIST, the practical perspective of industry, and the technical expertise of U.S. universities to address a nanotechnology research and development priority. It is one example of the creative methods the NNI uses to accelerate research that contributes to the Nation's economic competitiveness.

Beyond the Federal Program

In addition to coordinating and leading the Federal Government's nanotechnology R&D program, the NNI also engages with other governmental and nongovernmental organizations interested in the potential of nanotechnology to support U.S. competitiveness and promote the public good.

A number of regional organizations and state and local governments recognize the economic potential of nanotechnology and have begun to take steps to create nanotechnology-based clusters, that is, concentrations of universities, investors, and interrelated businesses with common manufacturing processes, customers, and suppliers. Because nanotechnology has the potential to impact almost every industry, whether high-tech or low-tech, there are opportunities for leveraging NNI R&D investments in every region. The NNI will support and foster these efforts by working with national, state, and regional groups, for example, to facilitate communication between local nanotechnology initiatives and to identify barriers to commercialization.

Goal 3: Develop and Sustain Educational Resources, a Skilled Workforce, and the Supporting Infrastructure and Tools to Advance Nanotechnology

A solid educational foundation, a skilled workforce, and a state-of-the-art R&D infrastructure are essential to the success of the NNI. Nanoscale science, engineering, and technology programs and other technology education programs and resources are required to produce a new generation of researchers and inventors working at the nanoscale. Educational programs continue to be developed with NNI support for all levels, including K-12 schools, community colleges, vocational schools, and major research universities. Building on this foundation, additional measures must be executed to develop and maintain a skilled nanotechnology workforce.

The supporting physical infrastructure, that is, highly specialized buildings and equipment, also is important. The NNI has created a network of interdisciplinary research centers and user facilities with modern equipment for nanometer-scale science and engineering research (see Figure 1, p. 12). Federal funding of state-of-the-art research facilities make these capital- intensive facilities accessible to researchers from all sectors. Additional facilities not specifically funded by the NNI, but which nevertheless support nanoscale science, engineering, and technology research, include synchrotron light sources and neutron-scattering facilities. In many instances, NNI user facilities and centers are located at, or near, these other facilities.

Nanotechnology Education

The NNI research centers, user facilities, and university- based research projects are designed and developed to foster multidisciplinary education, offer opportunities for teacher training, and stimulate the development of curricula and instructional materials. The NNI also provides hands-on training of technicians, undergraduates, graduate students, and postdoctoral researchers at universities, Federal laboratories, and other institutions.

In the future, the NNI member agencies will build on this investment by sustaining support for educational programs at all levels. Examples of specific efforts by the NNI member agencies to support and encourage the development, evaluation, and dissemination of educational materials for a broad cross-section of users and stakeholders include the following:

- Enhancing existing programs designed to develop middle school, high school, and undergraduate scientist–educators who can effectively introduce nanotechnology concepts into schools.
- Developing educational modules that incorporate nanotechnology into curricula for the many disciplines contributing to nanotechnology.
- Fostering international exchanges of students and researchers working in nanotechnology-related fields.
- Putting opportunities in place that bring together nanotechnology researchers from NNI-funded centers and user facilities to work with educational researchers and teachers, for example, through summer visiting research fellowships for teachers and undergraduate students.

The Nanoscale Science Research Centers Department of Energy User Facilities for Nanoscience Research

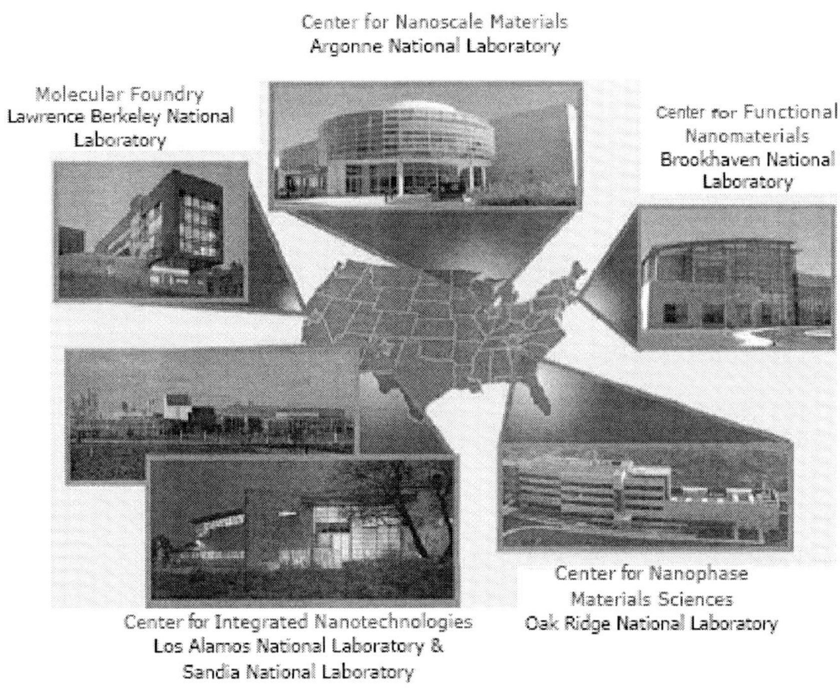

To accelerate synthesis, processing, fabrication, and analysis at the nanoscale, the Department of Energy's Office of Science has established five new Nanoscale Science Research Centers (NSRCs) since 2006. Each center houses particular expertise and instrumentation in a limited number of focused scientific themes. These centers are designed to be user facilities for interdisciplinary research at the nanoscale, serving as the basis for a national program that encompasses new science, new tools, and new computing capabilities.

User access is allocated based on peer merit review of submitted proposals.The NSRCs are each located in new, custom-designed space, and each supports a broad array of state- of-the-art capabilities, including sophisticated cleanrooms, high- power computational resources, laboratories for nanofabrication, and one-of-a-kind, signature instruments. They are co-located with preexisting Department of Energy user facilities. This allows the NSRCs and their users to take advantage of complementary capabilities at the national laboratories, including the neutron scattering centers, synchrotron light sources, and electron microscopy user facilities. This is one example of the U.S. nanotechnology infrastructure supported by NNI agencies.

Developing a Nanotechnology Workforce

A skilled workforce is essential to realizing the NNI vision. This workforce must include nanotechnology researchers, technicians, manufacturing engineers, and production workers. To develop this workforce, the NNI will build on the educational programs described above by promoting partnerships between industry, educators, and the Federally funded R&D system. Such partnerships will aim to provide access to training programs for careers in nanotechnology-related industrial sectors. Examples of activities the NNI will support and encourage include the following:

- Using career centers funded by the Department of Labor and other appropriate public venues to distribute information on nanotechnology and the career opportunities this field offers.
- Developing training programs that encourage workers to pursue nanotechnology-related career opportunities.
- Providing information on nanotechnology-related training opportunities on the NNI website.
- Assessing human resource issues, including workforce training, by comparisons with other countries through international benchmarking exercises.

Nanoscience Research for Defense and Dual-Use Technologies Department of Defense Facilities for Nanotechnology Research

Air Force, Army, and Navy research laboratories have developed unique and complementary approaches for fostering nanoscience research directed towards understanding and exploiting the unique properties that some engineered nanoscale materials exhibit. The

Air Force Research Laboratory Nanotechnology Initiative, the Army Research Laboratory Nanoelectronics Laboratory, and the Naval Research Laboratory Institute for Nanoscience have led this effort.

Each of the service laboratories maintains analytical and materials growth capabilities that make it possible to fabricate and test advanced nanotechnology-enabled devices. Each laboratory has different areas of specialization, ranging from microelectromechanical systems to biological and chemical sensors and from biologically inspired nanomaterials to ultra-high vacuum approaches for fabricating nanomaterials. This expertise covers a wide range of applications that are expected to be important to future defense technologies.

The Air Force Research Laboratory's Nanotechnology Initiative and associated programs have expanded the existing Air Force materials processing and characterization infrastructure. Research at these facilities accelerates the development of engineered nanoscale materials that address diverse aerospace challenges, including improved electro-optic sensors, composite materials for morphing vehicles, alternative energy generation and storage concepts, and improved propellants.

The Army Research Laboratory's nanoelectronics facilities provide world-class processing and characterization equipment for nanotechnology investigations, particularly for the development of nanoelectronic devices and test instruments. In particular, the Specialty Electronic Material and Sensors Cleanroom (SEMASC) is the Army's most extensive sensor prototyping laboratory, with a full suite of specialized etch, deposition, characterization, and photolithography tools. This user facility is open to all Army researchers, as well as to external university, ndustry, and government partners.

The Naval Research Laboratory's Institute for Nanoscience brings together scientists with disparate training and backgrounds to attack common goals at the intersection of their respective fields at the nanoscale. The institute's Nanoscience Research Laboratory includes a cleanroom with a full complement of advanced fabrication and metrology tools and vibration-isolated, controlled-environment laboratories for the most sensitive systems, such as transmission electron microscopes and scanning probe microscopes.

Full Utilization of the NNI Infrastructure

The extensive infrastructure established by the NNI over the past seven years includes centers and user facilities supporting research on nanomanufacturing and nanoscale characterization, synthesis, simulation, and modeling. This infrastructure is well suited to support the prototyping and demonstration stages of nanotechnology development. The NNI will support and encourage efforts to keep these facilities fully staffed and readily accessible to nanotechnology researchers from academia, industry, and government. The NNI agencies are committed to promoting broad access

to user facilities by all sectors, especially by small businesses. Future activities include development of an inventory of major tools and facilities, continued development of user training for these facilities, and efforts to publicize the availability of these resources. The NNI agencies will also evaluate these existing infrastructure and equipment investments, considering possible new needs for the long term. In the near term, however, the focus will be on maximizing the utility and utilization of the substantial infrastructure already in place.

Some specific activities that the NNI agencies will support in order to enhance utilization of NNI infrastructure include the following:

- Developing a means to increase awareness among the nanotechnology research community of the services that are available at the various user facilities and of opportunities for collaboration with researchers at the NNI research centers.
- Developing and disseminating a database of major equipment available at NNI user facilities.
- Providing expert staff to advise, collaborate, and, when appropriate, to operate or train users to operate available equipment.
- Upgrading instrumentation in order to maintain state-of-the-art facilities and equipment.
- Coordinating information exchanges among centers, networks, and user facilities funded by NNI member agencies; for example, information relating to practices for the safe handling of nanomaterials.

Goal 4: Support Responsible Development of Nanotechnology

Responsible development of nanotechnology entails research toward understanding the public health and safety and environmental implications of nanotechnology, as well as research toward promising, highly beneficial uses of the technology. Such an approach recognizes the value of supporting basic research to develop nanotechnology as well as research to address environmental, health, and safety concerns related to the use of the technology.

Responsible development of nanotechnology also entails establishing channels of communication with relevant stakeholders, in terms of both providing information and seeking input. Such communication allows the public and the NNI agencies to make well-informed decisions and builds trust among all stakeholders. Implicit in this approach is the recognition that public understanding and acceptance of nanotechnology and its applications are necessary components of successful commercialization.

The broad implications of nanotechnology for society can be grouped into two categories, namely (1) environmental, health, and safety (EHS) implications, and (2) societal dimensions. The NNI has made and will continue to make research in both these areas a priority. The implications of nanotechnology extend beyond borders, so international cooperation is an important part of the NNI strategy for responsible development in both of these categories.

Environmental, Health, & Safety Implications

The NNI's strategic approach to addressing the potential environmental, health, and safety implications of nanotechnology consists of comprehensive planning, research coordination, leveraging of forefront science through collaboration, and periodic review of EHS research needs and directions. This research strategy enables the NNI to identify new research needs and to pursue appropriate paths to meet those needs. Iterating the planning, research, and review cycle over the long term will provide the highest-quality information and data for government regulators, the public, and industry, helping to ensure that nanotechnology-enabled products are brought to the market responsibly.

The NNI agencies have funded high-priority, foundational nanotechnology EHS research since the inception of the initiative, and they continue to support a diverse and growing research portfolio. This broad portfolio is leveraged through the other investments of the various NNI agencies, along with the investments of industry and those of other countries.

The NNI has completed a first comprehensive assessment of EHS research and information needs and ongoing activities. Government subject matter experts led this assessment, informed by inputs from program managers, funding decision makers, and the public. Additional input was gained through U.S. participation in parallel international efforts. The process involved research, regulatory, and other Federal agencies, and thus promoted communication of needs, capabilities, and priorities between those communities. This activity provides the basis for current NNI planning and guidance activities in this area.

The information collected during the process just described was used to establish EHS research needs for engineered nanoscale materials, and a priorities document has been published. Documents such as this provide guidance to those who implement necessary research, that is, Federal agencies, industry, academia, and other relevant groups. A document that describes the NNI strategy for addressing the identified priorities for nanotechnology-related EHS research is in preparation.

The NNI plans to conduct periodic reviews of the needs, priorities, and strategies for nanotechnology EHS research. These reviews will establish whether new developments or discoveries require that research directions and goals should be changed or redirected. This approach to planning and implementing continuing nanotechnology EHS research will enable the NNI to adapt to the dynamic aspects of research and discovery and to pursue the most promising paths for addressing newly identified research needs. This approach takes advantage of the full depth of the Federal Government's resources and expertise and minimizes the pursuit of unnecessary and duplicative paths of research.

Societal Dimensions

Nanotechnology, like other new areas of technology, will impact society in ways that may be difficult to predict. The NNI supports ongoing research pertaining to ethical, legal, and societal implications (ELSI) of nanotechnology, in order to better understand its societal ramifications, to encourage the distribution and exchange of insights from leading experts in this area, and to develop avenues for societal input into nanotechnology development. As part of this effort,

NSF supports two research centers focused on the study of the societal dimensions of nanotechnology.

The NNI also promotes public outreach, engagement, and communication of research findings, including those related to understanding societal dimensions of nanotechnology. An important tool that has been and will continue to be used in this effort is the NNI website, www.nano.gov. This is a major vehicle for providing information ranging from research news to funding opportunities. The website is linked to sites that host educational resources, safety databases, and other nanotechnology-related information. There will also be continued support for efforts to educate the public through means such as those currently led by the NSF-funded Nanoscale Informal Science Education (NISE) Network, which is a combination of exhibits and resources aimed at educating the public about nanotechnology. Another example of public education and outreach is the National Institute of Environmental Health Sciences' Nanotechnology Webinar series. This program brings the public, industrial hygienists, and public health advocates into a web-based dialogue with nanotechnology subject matter experts. Other outreach efforts, such as media roundtables, will also continue, and the NNI will explore ways of building capacity for public engagement, using vehicles such as web dialogues.

NNI Agencies Leveraging Resources Joint Funding of Research on Potential EHS Implications

The National Nanotechnology Initiative (NNI) supports collaborations among multidisciplinary research groups and across agencies. NNI agencies jointly fund research on potential environmental, health, and safety implications of nanotechnology. Below are examples of how the agencies of the NNI share resources and work together to achieve common objectives.

- The Environmental Protection Agency (EPA) initiated, with subsequent support from the Department of Energy (DOE) and the National Science Foundation (NSF), funding for research that focuses on investigating the environmental and biological fate, transport, and transformation and exposure/bioavailability of engineered nanoscale materials. This research program encourages international collaborations, which are critical for an integrated and comprehensive approach toward understanding the impact that engineered nanomaterials may have on humans and ecosystems. This EPA-led program began in fiscal year 2005; participation of other agencies has varied from year-to-year.
- The Army Engineer Research and Development Center is conducting research that focuses on the characterization, fate and transport, toxicology, and modeling of engineered nanoscale materials specifically related to militarily relevant nanotechnologies. The goal of this research is to provide technology developers with tools for "smart" development and use of engineered nanomaterials. This research has strong connections to other Federal agencies, including the Air Force, EPA, U.S. Geological Survey, and the National Institute for Occupational Safety and Health (NIOSH). Funding began in fiscal year 2006.
- The National Institute of Environmental Health Sciences (NIEHS) of the National Institutes of Health (NIH) is partnering with EPA and NIOSH, as well as with five other institutes at NIH, to support research that investigates the systemic, cellular, and molecular responses to manufactured nanoscale materials. This program began in fiscal year 2007.
- NSF and EPA have issued a joint funding solicitation for proposals to create a Center for Environmental Implications of Nanotechnology. This center will conduct research and education on the implications of nanotechnology for the environment and living systems at all scales. The center will address interactions of naturally derived, incidental, and engineered nanoscale and nanostructured materials, devices, and systems with the living world. Funding is planned to begin in fiscal year 2008.

High-Impact Application Opportunities

The power of nanotechnology is rooted in its potential to transform and revolutionize multTiple technology and industry sectors, including aerospace, agriculture, biotechnology, homeland security and national defense, energy, environmental improvement, information technology, medicine, and transportation. Discovery in some of these areas has advanced to the point where it is now possible to identify high-impact application opportunities for nanotechnology and areas where critical research will significantly advance these potential applications.

Specific examples of these opportunities and needs are given on the pages that follow this introduction. These examples illustrate a few of the potential benefits that nanotechnology offers. They are of interest to a wide variety of NNI member agencies. These examples are derived in part from ideas generated at NNI-sponsored topical workshops, at which experts from academia, industry, and government examined a particular aspect of nanotechnology in detail (see Appendix A for a list of these workshops).

Each example lists interested NNI agencies. Many of the NNI agencies that fund nanotechnology R&D have supported, and will continue to support, the foundational scientific research and technology development that will ultimately meet the research needs and lead to the applications described here. One or two agencies often have a particularly strong connection to a given application area. Agencies with mission-driven needs pursue nanotechnology applications that address their individual requirements. In many cases, however, multiple agencies have an interest in an application, be it from a research, regulatory, or policy perspective. In some cases, coordinated research across agencies is necessary in order to realize an application opportunity or meet a research need. The relationships between the agencies and selected application areas are summarized in Table 3.

The application and research needs discussed here are visionary, yet realistic. The time horizon for realizing these goals is long-range, ten years or more. Because scientific research and product development is never formulaic, it is likely—indeed, it is probable—that research and discovery will lead in unanticipated directions. Transforming discovery into products is a complex process for any technology and typically takes decades; this will be no different for nanotechnology.

The most effective means to achieve the potential of the nanotechnology application opportunities discussed here will be to pursue a variety of technical approaches. Similarly, a wide breadth of collaboration often will be required.

The high-impact application opportunities given on the following pages, although not comprehensive, are examples of the important national needs and major industry sectors on which nanotechnology could have an impact. Equally, the critical needs are only some of the important advances required to move nanotechnology to its next phases. It must be emphasized that this is just an illustrative sampling. Other important nanotechnology R&D priorities will—and must—be pursued.

Table 3. NNI Agency Contributions To Selected Nanotechnology Application Areas

● – Central Role ○ – Supporting Role

Application Area / Agency*	CPSC	CSREES	DOD	DOE	DHS	DOJ	DOL	DOT	EPA	FDA	FS	IARPA	ITC	NASA	NIH	NIOSH	NIST	NSF
Aerospace			●	○	○			○				○	○	●			○	○
Agriculture & Food		●	○	○	○	○			○	●	○							○
National Defense & Homeland Security		○	●	○	●	○			○	○		●	○	○	○	○	○	○
Energy		○	●	●	○				○		○	○	○	○			○	○
Environmental Applications	○	○	○	○	○		●		●	○	○				○	○	○	●
Information Technologies			●	○	○							○	○	○	○		○	●
Medicine & Health	○	○	○	○			●		●	●				○	●	○	○	○
Transportation & Civil Infrastructure			○	○	○			●	○		○						○	○

EARLY DETECTION OF LIFE-THREATENING DISEASES

Rapid, Multiplexed Detection of Disease Markers

Interested Agencies: CSREES, DOD, FDA, NIH, NIST, NSF

Early detection of a life threatening disease, such as a cancer, or rapid and accurate detection of a life- threatening exposure, such as to a chemical or biological agent, can make a difference between death and survival. Similar technological approaches can be used to detect the molecular signatures of disease, an environmental exposure, or the presence of toxins or microorganisms in food. However, current detection methods often require multiple steps, do not provide data in real time, are not quantitative, and have limited sensitivity and specificity. Nanotechnology has the potential to overcome the limitations of current approaches and thereby advance the diagnosis and treatment of life-threatening diseases, enhance our national security, and improve food safety.

To make rapid, accurate, real-time detection possible, a number of advances will be required. The molecular signatures of disease, also called biomarkers, and biological threats are often present at concentrations that are too low to be measured by current technology, so new devices will need improved sensitivity. Additionally, many sensors are not able to measure multiple parameters simultaneously. This multiplexing capability is needed to distinguish one disease or exposure from another and will improve the specificity and accuracy of the disease or exposure identification. Ideally, these devices should be capable of measuring large numbers of markers simultaneously on a single, easily obtainable sample, such as breath, sweat, or saliva, be portable, and be easy to use in a doctor's office or by a first responder.

One example of a promising new medical approach that uses bionanotechnology is the bio-barcode assay for the detection of disease-related proteins or DNA in tissue samples. In its first application, the bio-barcode assay uses gold nanoparticles to amplify and detect amyloid beta-derived diffusible ligands (ADDLs), a molecular signature for early-stage Alzheimer's disease. This method is a million times more sensitive than the current point-of-care diagnostic tests, specific for molecular signatures of disease, and quantitative.

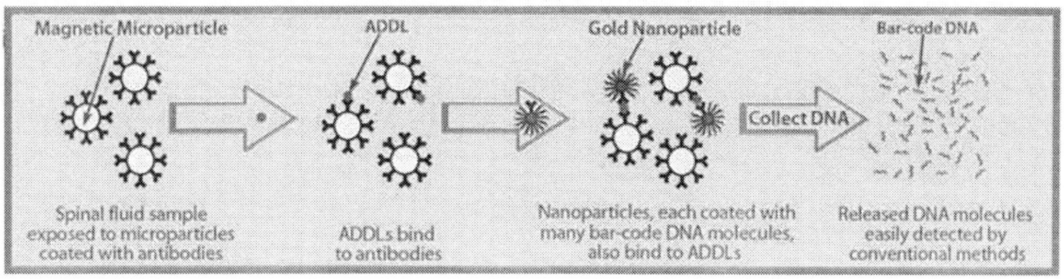

Schematic presentation of the bio-barcode amplification and magnetic collection process for detecting the presence of ADDL, a molecular signature of early-stage Alzheimer's disease, using gold nanoparticles coated with DNA. Adapted from Proc. Natl. Acad. Sci. USA; original image © 2005 National Academy of Sciences, all rights reserved.

Other promising medical diagnostics using nanotechnology include methods to detect infectious agents and improved diagnostic imaging systems to examine tissues and organs for disease diagnosis, such as computed tomography and magnetic resonance imaging. Looking further down the road, many of these imaging agents could be coupled with targeted drug delivery.

Knowing It's Safe Exposure Measurements For Engineered Nanoscale Materials Interested Agencies: CPSC, DOD, DOE, DOL, DOT, EPA, FDA, NIH, NIOSH, NIST, NSF, USGS

Knowing the concentration and types of engineered nanoscale materials in the air, in water, and on surfaces is essential to evaluating the potential impacts on workers, consumers, the public, and the environment.

Definitive methods for identifying engineered nanoscale materials in the environment are neither well developed nor readily available. Presently, estimating exposure to and concentrations of nanomaterials is possible only for a limited number of material types. The results of these measurements are often inaccurate, and the process is cumbersome and typically involves collecting samples in the field for subsequent analysis in the lab.The instruments used in these analyses incorporate sophisticated technologies and methods, such as electron energy-loss spectrometry, that require expert operators to collect the data and interpret the results. Moreover, a single instrument may cost hundreds of thousands of dollars and be capable of measuring only one or two of the many parameters of interest, for example average size or elemental composition. Even these limited measurements can take hours, if not days, to complete.

For exposure and concentration measurements of engineered nanoscale materials to become feasible, existing methods for measuring the relevant parameters must be simplified and refined and new approaches must be developed. For these measurements to become practical, the instruments will need to be affordable, portable, and work in the field, and provide exposure and concentration information in real time. Ideally, these instruments would also measure several relevant parameters, such as size distribution, shape, charge state, and chemical composition. Such instrumentation would greatly facilitate risk assessment analyses and would facilitate the development of appropriate exposure mitigation programs.

Upper: Nanoaerosol mass spectrometer illustrating the size (footprint of several square feet) and complexity of current instrumentation for measuring nanoparticles in the atmosphere (courtesy of M. V. Johnston, University of Delaware). Lower left: Concept of a future-generation instrument (courtesy of Sandia National Laboratory).

Nanotechnology itself could potentially provide platforms for a next generation of analytical instruments, perhaps through the development of high-density, multianalyte sensor arrays.

LETTING NATURE DO THE WORK

NANOBIOTECHNOLOGY INTERESTED AGENCIES: CSREES, DOD, DOE, FDA, NASA, NIH, NSF

A key challenge for nanotechnology is the fabrication and assembly of nanometer-scale materials, devices, and systems. Biology performs these functions in living cells, which in many respects act as integrated nanomanufacturing factories. A cell includes nanomaterials (proteins, membranes, etc.), nanodevices (motor proteins, ion channels, etc.), and even functional nanosystems (e.g., mitochondria) that are created from individual molecules by hierarchical self-assembly. Biological capabilities can be harnessed directly, borrowed, or taken as inspiration for future nanomanufacturing.

Naturally occurring nanostructured organisms such as diatoms have been converted from silica into a range of ceramic nanomaterials with new properties. Cells and viruses can also be engineered to manufacture or assemble nanomaterials. Certain bacteria naturally produce magnetic nanoparticles and have already been engineered to produce nanoparticles coated with specific proteins.

An image, reconstructed from transmission electron micrographs, of a biologically inspired complex composed of a semiconductor quantum dot encapsulated in a protein shell.Reproduced by permission from Nano Letters, © 2006 American Chemical Society.

Researchers have used biological methods to discover viruses that can be used as scaffolds for selective attachment and growth of semiconductor nanoparticles, and have produced the first virus-assembled nanoelectrode and virus-assembled battery. Scientists are also learning from processes organisms use to create complex inorganic nanostructures, including those with unique chemical, optical, and mechanical properties.Progress in genomics, proteomics, and bioinformatics is rapidly expanding our understanding of how the genetic code enables cells to turn molecules into nanostructures and nanosystems. Harnessing this biological knowledge will enable flexible, scalable, and affordable nanomanufacturing for the pharmaceutical, biochemical, nanoelectronics, and materials industries.

SAFE AND AFFORDABLE WATER
NANOTECHNOLOGY-BASED WATER PURIFICATION AND TESTING
INTERESTED AGENCIES: CSRESS, DOD, DOE, DOS, EPA, NASA, NIH, NSF, USGS

Clean water is a basic necessity for people and animals and vital to agriculture. The worldwide supply of potable water is limited and demand is increasing. Disease-causing organisms often contaminate water in less developed parts of the world. Gasoline additives, dry cleaning chemicals, and agricultural pesticides sometimes foul the water supply in industrialized countries. Water pumped from deep aquifers often contains high levels of contaminants such as iron, manganese, and even arsenic, or has high salinity.

Many different technologies already exist for removing various contaminants from water, including distillation, chlorination, ultraviolet irradiation, and selective membrane filtration. Each has its drawbacks.

Most are energy-intensive. Some generate byproducts that have unfavorable environmental and health impacts.

Nanotechnology could help meet the need for safe, affordable water through inexpensive water purification and rapid, low-cost detection of impurities. Solutions are likely to come from many directions. Some may be applied locally while others will be best suited to central treatment facilities. Nanostructured mixed metal oxides can destroy biological toxins in water through photocatalysis and oxidation. Coatings made from nanomaterials minimize biofilm fouling, a problem with existing membrane technologies.

Progress has already been made. Experiments with palladium-doped iron nanoparticles have demonstrated the capability to remove organic contaminants from groundwater. Magnetite nanoparticles have shown promise for removing arsenic from drinking water. Sensors based on nanostructured surfaces have been used to detect metals, biotoxins, and organic compounds accurately and rapidly. These sensors also could be incorporated into sophisticated control systems for sensing and then treating specific contaminants.

Nanotechnology-based membrane materials, impurity-scavenging nanoparticles, and portable devices for on-site detection and diagnosis of contaminants are priorities in water purification research. It will be important to investigate the scale-up of low-impact manufacturing processes for these new systems. Finally, it will be important to ensure that there are no unwanted environmental and health impacts of engineered nanomaterials used in the new purification processes.

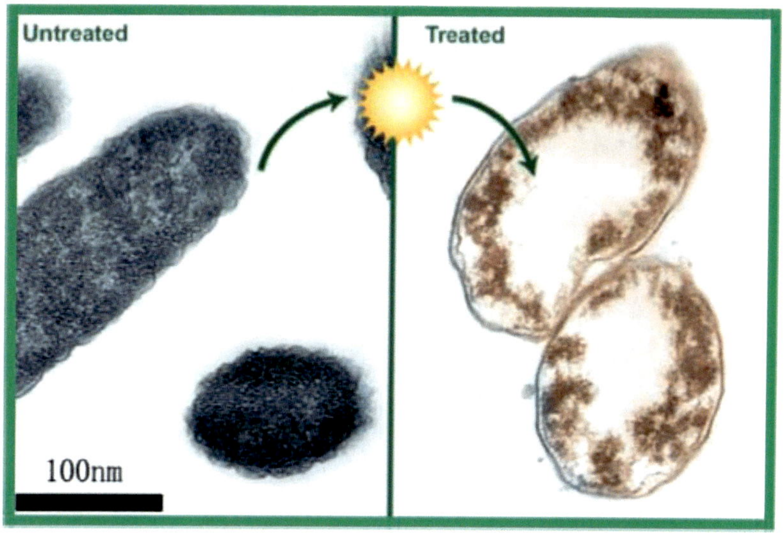

Transmission electron microscope micrograph showing the destruction of E. coli bacteria in water using visible light, a process made possible by the presence of nanoscale titanium oxide (courtesy of J-K. Shang and B. M. Pianfetti, University of Illinois at Urbana-Champaign).

SMARTER COMPUTERS
FUTURE INFORMATION PROCESSING TECHNOLOGY FROM NANOTECHNOLOGYINTERESTED AGENCIES:
DOD, DOE, DOT, FDA, NASA, NIH, NIST, NSF

Computers and other advanced electronic systems are at the foundation of our Nation's technological and economic strength. Complex natural phenomena such as climate patterns are modeled using advanced computers. Airplanes and car engines are designed with sophisticated information technology systems. Three-dimensional diagnostic images of human organs are constructed by computers coupled to imaging technologies.

Complementary metal-oxide semiconductor field-effect transistor (CMOS-FET) technology is the engine that propels modern computers and electronics. Historically, CMOS-FETs have gotten smaller, year after year. This has meant that more components can befit onto a computer chip, one reason why computers and other electronic systems have gotten ever more powerful over time. As individual circuit components approach nanoscale dimensions, however, conventional electronic devices become unreliable and waste too much power, both of which will eventually limit the ability to miniaturize.

Nanotechnology offers new approaches for continuing improvements in information technology. One of the most anticipated developments will be discovery of a replacement for the logic switch around which today's computers are designed, that is, the CMOS-FET. Ideally, future nanotechnology-enabled logic switches will provide increased performance while using significantly less electrical power, will be integrable with existing fabrication processes, and will lead to computers with much higher component densities.

Scanning electron microscope images of a nanowire crossbarmemory circuit, in which the memory elements are made fromoverlaying silicon and titanium nanowires. This circuit has amemory element density of 100 billion bits per square centimeter.Lower right: Image of entire circuit. Upper left: Magnified imageof nanowire crossbar memory junctions. Wire-to-wire distance isapproximately 33 nm. Reproduced by permission fromNature, © 2007 Macmillan Publishers, Ltd.

Controlling the structure of matter at the nanoscale will enable future electronic devices that will operate more efficiently at much lower power than existing devices and systems. Graphene, a flat sheet-like molecule made of carbon atoms, offers an example. Electrons can travel great distances in sheets of graphene, resulting in less waste heat.

Moreover, depending on how a graphene sheet is oriented, it can act either as a conductor or a semiconductor, two essential components in an electronic circuit. These characteristics offer the potential advantage of being able to create energy-efficient, multicomponent devices from a single sheet of graphene. There are numerous other areas where nanotechnology can contribute to the continued improvement of information technology systems. These include the use of nanostructured materials for better thermal management and superior dielectric properties.

Nanotoxicology: Doing it by the Numbers Predicting Toxicity Before Manufacturing Interested Agencies: CPSC, DOD, DOE, DOL, DOT, EPA, FDA, NIOSH, NIH, NIST, NSF

The behavior of engineered nanoscale materials in biological systems is directly related to the physical and chemical properties of that material. The rapidly increasing number of nanomaterials in development and manufacture, as well as the exquisite sensitivity of a material to its biological microenvironment, makes it difficult to predict biocompatibility or toxicity in humans and the environment. Currently, decisions to proceed to manufacture or to release products for commercial use are based on laboratory assessments of toxicity.

Conventional toxicity assessment often employs a tiered system of assays. A material that gives positive indications of toxicity in the first-tier assessment is then analyzed more rigorously in a dose- response model. These data, plus exposure measurements from the exposure site, are analyzed, and an estimate of risk is calculated. These analyses are frequently time-, nanomaterial-, and equipment-intensive. Computational models that predict biological and environmental response of a new material based on knowledge of structurally similar materials have been developed for several classes of chemicals.

Development of similar predictive models for nanomaterials will require a thorough understanding of the biological interaction of a few well-characterized standard nanomaterials in well-characterized biological and environmental systems. These data would be used to define basic structure-activity relationships and would serve as the basis for creating models that predict the biological and environmental response of new or more complex nanomaterials. Development of the computational components of a predictive model would occur in parallel with the biology studies and require translation of the models for chemicals to nanoparticles and complex three-dimensional nanoscale structures.

The ability to predict biological and environmental response to nanomaterials through a computational model and to adjust physical and chemical parameters of the materials early in product research and I development has the potential to shorten the time for toxicity testing from several weeks to a matter of hours. It will also provide a labor and cost benefit to the manufacturer, new tools for risk assessment by p regulatory agencies, and protection of humans and the environment from harmful exposures.

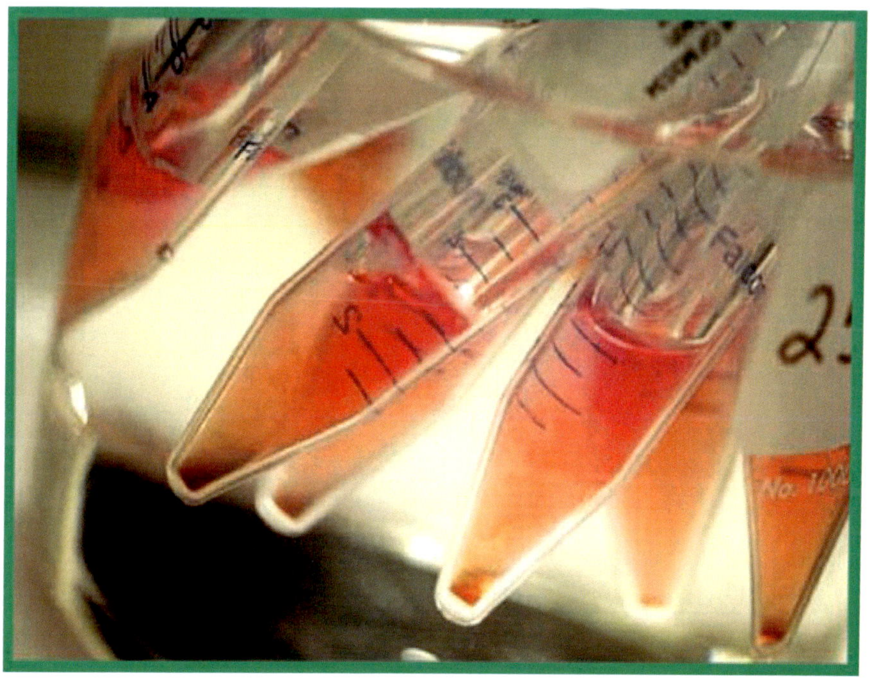

Vials of quantum dots and blood plasma following centrifugation, part of a study of the effects of quantum dots on immune cell function. Such studies are used to evaluate the safety of nanomaterials in humans. Computational methods could augment or replace these time-consuming and expensive laboratory tests. (Image courtesy of NCL).

Could, Would, or Should? Societal Dimensions of Nanoscale Science and Technology Interested Agencies: CPSC, CSREES, DOE, DOL, FDA, NASA, NIH, NSF

Nanotechnology is a science in motion. Anticipating the societal impact of nanotechnology requires balancing uncritical expectations and fears with values and issues raised by ethical, legal, and societal discussions regarding its development and applications.

Potential applications of nanotechnology are expected in a wide range of fields, from agriculture and food, biomedical applications, electronics, energy, environmental remediation, and material sciences to computing and other technologies. However, uncertainty surrounding some future applications and their various potential impacts could hamper development of nanotechnology.

The research challenge will be to identify and, to the extent possible, quantify potential societal developments (social, economic, workforce, educational, ethical, and legal) that could result from discoveries in nanotechnology. Research on the processes of innovation, diffusion, and adjustment will generate information on potential societal impacts.

Additionally, modeling tools, such as scenario analysis and multiagent modeling, can be used to assess potential impacts and resulting societal interactions. This provides information concerning potential impacts to policy makers, business, and the public, allowing for the development of beneficial applications of nanotechnology, while reducing and managing any potential unintended consequences. Scenario analysis is effective in identifying issues and researching hypotheses and thus can be an effective tool for theoretical analysis, while multiagent modeling is akin to scenario analysis, but is carried out through computer simulation.

It could prove interesting to compare the history of nanotechnology with the history of other technologies, such as the Internet, genetically modified organisms, aircraft, stem cells, and computers, to draw lessons that might be relevant. Alternative histories, for instance, can be built using such familiar technologies as the automobile. How might today's society approach its introduction were it being developed now? How would society attempt to avoid unintended impacts?

More Energy-Efficient Transportation

Interested Agencies: DOD, DOE, DOT, EPA, NASA, NIH, NIST, NSF

Lightweight Magnetic and Structural Nanomaterials

A study by the National Academies, titled Effectiveness and Impact of Corporate Average Fuel Economy (CAFE) Standards, found that reducing the weight of a vehicle by twenty percent results in a fifteen percent reduction in fuel consumption. A few specific examples illustrate the importance and magnitude of this relationship between vehicle weight and transportation fuel efficiency. AirTran recently installed carbon-fiber-based seats in its 737-700 Series aircraft, which shaved 19.4 pounds from each row of seating and is estimated to result in fuel savings of $2,000 per year per aircraft. Boeing estimates that weight savings due to use of carbon fiber composite materials in its new 787 aircraft will increase fuel efficiency by as much as twenty percent. Such reduction in fuel consumption for transportation has the additional benefit of reduced outputs of greenhouse gases.

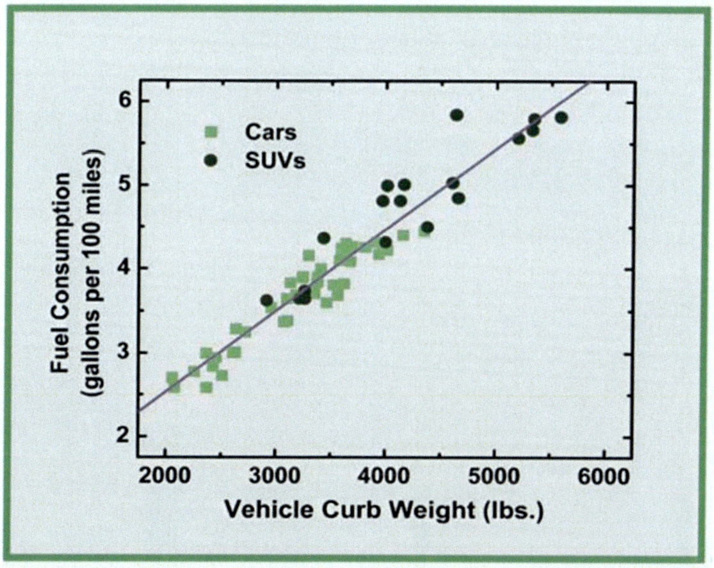

Relationship of fuel consumption rate to vehicle weight for a representative set of vehicles used in the United States. Data from: Effectiveness and impact of Corporate Average Fuel Economy (CAFE) Standards, National Academy Press (Washington, DC: 2002).

Nanotechnology offer possibilities for production of new strong and lightweight materials that also mayexhibit superior magnetic and mechanical properties. These materials could result in weight reductionin transportation vehicles. For example, calculations predict that nanocomposite magnets could havemagnetic strengths twice as highas those of the best commercialmagnets in use today. Thesemagnetic materials could significantlyreduce electric motor weight andsize in hybrid automobiles, railwaysystems, and subways. Application ofnanostructured composite materialsmay also allow designers to reducethe weight of structural componentswithout compromising stiffness orother properties that are essential

Torealizethe potential offor passenger safety.nanotechnology-based materials, key scientific and technological barriers must be overcome. Improved synthesis and processing techniques must be developed to enable bulk fabrication of nanomaterials, to ensure high quality and purity of the nanomaterials, and to control their structure at the nanoscale.

BETTER MEASUREMENTS
REFERENCE MATERIALS FOR COMMERCE AND SAFETY MONITORING INTERESTED AGENCIES: CPSC, EPA, DOD, DOL, FDA, NIH, NIOSH, NIST, USGS

The chemical composition and physical characteristics of today's engineered nanoscale materials can vary substantially between batches and from vendor to vendor. Products manufactured from such variable starting materials have inconsistent performance characteristics. Health and safety testing studies yield seemingly variable results. Accurate calibration of environmental monitoring systems is difficult.

A suite of reference standards (stable materials with rigorously characterized chemical and physical properties) could help to eliminate these problems. Reference standards are used in laboratories to perform instrument calibrations, by manufacturers to verify the composition or properties of their products, and in clinical laboratories to validate the accuracy of specific measurements. Material standards with property values that have been determined using established and definitive methods are essential in the validation of new measurement methods.

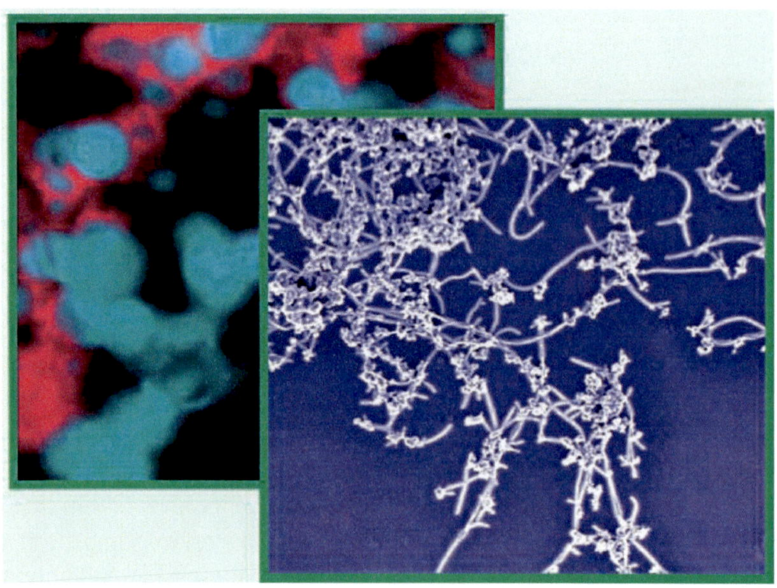

Lower right: Electron micrograph of a carbon nanotube sample showing how the sample contains significant impurities such as metallic particles and other forms of carbon. Field width is approximately 4 lam. Upper left: Electron energy loss spectrometry reveal metallic impurities (pink) and along with carbon (light blue). Field width is approximately 100 nm. (Images courtesy of NIST.)

Standards are needed for engineered nanoscale materials. Different groups or classes of standards will be needed by different sectors. For example, materials necessary for performance evaluations in manufacturing may consist of nanocomposites or nanoclays relevant to a particular industry. Materials needed for toxicology and ecotoxicology research include titanium dioxide, zinc oxide, and carbon nanotubes. However, significant R&D is needed on many fronts. Pure samples are needed. The critical and relevant physical parameters (size, shape, composition, etc.) must be identified. Reliable methods of distributing these standards must be established. Instruments with the ability to determine the elemental composition, location, and chemical state of all atoms in nanomaterials must be developed.

Creating a suite of standards for engineered nanoscale materials will require a coordinated interagency research effort. Multiple NNI agencies will need to be involved in developing and validating instrumentation and methods. Agencies with needs to measure specific materials will play a role in selecting material for which standards will be developed. The research-sponsoring agencies will be critical to funding studies that identify relevant chemical and properties for standardization. Regulatory agencies will have an essential role in describing the use of these standards, for example in environmental and workplace monitoring settings.

Energy Security
Meeting the Energy Challenge
Interested Agencies: CSREES, DOE, DOD, NASA, NSF

The world demand for energy is expected to double by the year 2050. The difficulty of meeting this demand is compounded by the growing need to protect our environment. To resolve this challenge and provide secure energy resources, energy efficiency must be increased and new "clean" energy sources developed.

Nanotechnology may provide novel solutions to these challenges. At the root of the opportunities nanoscience offers in energy research is a simple fact: many of the elementary steps in the storage and conversion of energy take place on the nanoscale. Furthermore, new functionalities of nanoscale materials emerge from high surface area, quantum size effects, high aspect ratio, reduced dimensionality, and intimate connectivity. The properties of engineered nanoscale materials, as well as the techniques used to image, measure, and manipulate these materials, create new opportunities for energy technologies.

Nanomaterials offer new ways to store and transform energy and to control catalytic activity. Photovoltaic cells made of nanoscale semiconductors offer several mechanisms to enhance the efficiency and reduce the cost of solar energy. For example, nanomaterials can be designed to absorb a broad range of colors from the solar spectrum. Harvesting solar energy with twenty percent power conversion efficiency and a hundred times lower costs than present photovoltaic technology could be within reach. Solid-state lighting involving the use of semiconductor-nanocrystalbased light emitting diodes for general illumination is another rapidly developing technology with the potential to save immense quantities of energy. Nanostructured materials offer a number of opportunities for the development of enhanced power storage and conversion, and for the design of improved catalysts. Nanostructured catalysts can achieve high selectivity and turnover. For example, it is envisioned that the emerging tools of nanoscience will enable one-step conversion of cellulose or carbon dioxide into liquid fuels and economical photocatalytic conversion of water into hydrogen and oxygen. Fuel cell catalysts fabricated from nanoparticles of ruthenium-platinum have been shown to exhibit enhanced resistance to poisoning. Three- dimensional structuring of electrode materials can greatly enhance power storage in batteries and ultracapacitors, and provide advantages over existing technologies in terms of power losses and charge rates. Development and commercialization of these technologies would usher in entirely new energy storage systems that change the economics of solar and wind technologies, greatly contributing to energy independence and security.

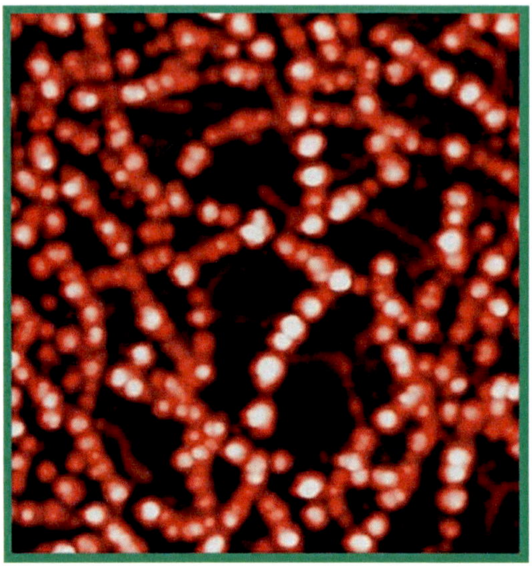

Atomic force microscope image of palladium nanoparticles electrodeposited on the surfaces of single-walled carbon nanotube, as part of a study on using these materials as mechanically flexible hydrogen sensors. Reproduced by permission from Appl. Phys Lett. © 2007, American Institute of Physics.

Coordination and Assessment Activities of the NNI

The Nanoscale Science, Engineering, and Technology (NSET) Subcommittee leads the interagency coordination of the Federal Government's nanotechnology R&D enterprise. The NSET Subcommittee is under the National Science and Technology Council's (NSTC) Committee on Technology (CT). The participating agencies of the NNI execute the research, development, and funding functions of the NNI. External bodies periodically review the progress of the NNI and recommend strategies for further strengthening its activities. Figure 2 shows the various entities that play a role in NNI coordination and assessment activities and their relationships to each other. The roles of the various entities and their interactions are further described below.

BACKGROUND

The beginnings of the NNI can be traced to September 1998, when the Interagency Working Group on Nanotechnology was formed. This working group sponsored workshops and studies to define the state of the art in nanoscale science and technology and to forecast future developments. Two years later, this effort was raised to the level of a Federal initiative. The NSET Subcommittee was then established under the NSTC's Committee on Technology to serve as the body responsible for interagency coordination. The National Nanotechnology Coordination Office (NNCO) was subsequently established as the point of contact on Federal nanotechnology R&D activities and to provide technical and administrative assistance to the NSET Subcommittee.

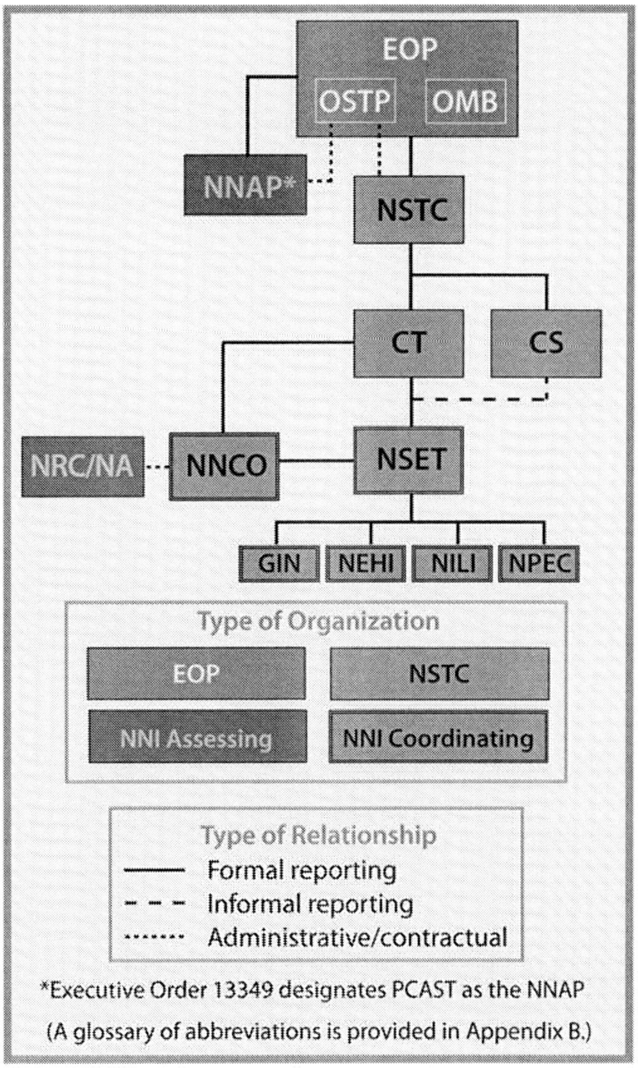

Figure 2. Entities Involved in the Coordination and Assessment of the NNI.

The importance of a coordinated Federal program for nanotechnology R&D was further recognized with the enactment of the 21" Century Nanotechnology Research and Development Act in December 2003 (hereafter, "the Act"). This legislation authorized appropriations for nanotechnology research, formalized many of the coordination structures that the NSTC had organized, and established additional mechanisms to ensure that the Government developed sound, informed nanotechnology R&D strategies and policies. This legislation also created the National Nanotechnology Advisory Panel (NNAP), called for a triennial review of the NNI by the National Research Council of the National Academies (NRC/NA), and established specific functions for the NNCO.

NSET Subcommittee

The NSET Subcommittee performs many of the coordination activities of the NNI, such as developing the NNI Strategic Plan, preparing the NNI supplement to the Presidential Budget, and sponsoring workshops that inform the Government's nanotechnology-related decision-making processes. Each agency participating in the NNI is represented on the NSET Subcommittee; a list is given at the front of this chapter. A Co-Chair from the Office of Science and Technology Policy (OSTP) and a Co-Chair from an NNI agency lead the NSET Subcommittee. It meets at least six times each year.

The member agencies of the NNI, working through the NSET Subcommittee, establish broad goals, identify priorities, and formulate plans for the NNI. This high-level framework, in turn, guides and informs the member agencies in developing their nanotechnology R&D implementation plans. The NSET Subcommittee also plans and sponsors interagency activities aimed at achieving the NNI goals. The subcommittee promotes a balanced investment across all of the agencies, to address the critical elements needed to support the development and utilization of nanotechnology. The subcommittee interacts with pertinent academic, industry, state and local government groups, and international organizations. The NSET Subcommittee also keeps the NSTC's Committee on Science (CS) informed of its activities.

Working Groups of the NSET Subcommittee

Four working groups currently exist under the NSET Subcommittee. These subgroups are focused on areas that the subcommittee recognizes will benefit from focused interagency attention and activity.

Global Issues in Nanotechnology Working Group

The NSET Subcommittee's Global Issues in Nanotechnology (GIN) Working Group coordinates international activities in nanotechnology, monitors foreign nanotechnology programs, and seeks to broaden international cooperation and communications with respect to nanotechnology R&D. The working group has representatives from all Federal agencies that have active nanotechnology R&D programs as well as from numerous agencies that have oversight roles in international affairs.

NNI agencies and representatives participate in many international activities, including bilateral and multilateral cooperative programs, monitoring of foreign nanotechnology R&D, and promotion of the trade and commercial interests of the United States. Cooperation and collaboration with other nations on nanoscale science and technology R&D, particularly in pre-competitive and non-competitive research areas, can further the progress of the NNI while helping our international partners achieve their own goals. The United States seeks to foster mutually beneficial relationships with other countries, in order to establish a framework for the safe, secure, and responsible use of nanotechnology worldwide.

In keeping with the NNI goal of supporting responsible development of nanotechnology, international cooperation is important to ensuring that health, environmental, safety, and security concerns surrounding the use of nanotechnology-enabled products are appropriately addressed by the global scientific community and relevant regulatory agencies. Effective communication among scientists, regulators, policy makers, consumers, industrial leaders, and other stakeholders also will be enhanced by cooperation with international partners. The development of a healthy global marketplace for nanotechnology products and ideas will require the establishment of consumer confidence, common approaches to nanotechnology environmental, health, and safety issues, efficient and effective regulatory schemes, and equitable trade practices for nanotechnology, not just in the United States, but worldwide.

Nanotechnology Environmental and Health Implications Working Group

Nanotechnology holds great potential. But in order to realize this potential fully, it is important to know how to handle and use engineered nanoscale materials safely in the lab, in manufacturing, in the workplace, and in consumer products, and to understand possible environmental, health, and safety impacts. NNI activities provide channels for the exchange of information and the formulation of research strategies that accelerate progress towards understanding the risks and benefits of nanotechnology and that lead to practices that protect human health and safety as well as the environment. The NSET Subcommittee's Nanotechnology Environmental and Health Implications (NEHI) Working Group coordinates many of the NNI activities in this area.

The NSET Subcommittee and NEHI Working Group provide leadership in establishing the national nanotechnology environmental, health, and safety research agenda and in communicating data and information related to environmental and health aspects of nanotechnology between NNI agencies and with the public. New tools and methods are required for the research that will enable risk analysis and assist in regulatory decision-making, and NNI activities support development of these tools and methods. All these activities require nanotechnology standards, including standard nomenclature and terminology, and the NNI works with consensus-based standards organizations to develop standards for nanotechnology. The outcome of this research, data collection, and tool and standards development is the information that provides the basis for the drafting of guidance in the safe handling and use of nanotechnology-derived materials and products by researchers, workers, and consumers.

The combined efforts of many sectors of the R&D community are required to address the environmental, health, and safety research priorities to strengthen the scientific foundation of risk assessments and risk management of nanotechnology-enabled products. The NNI agencies engage and interact with regional, state, and local groups, educational institutions, industry- supported groups, and non-governmental organizations. Interactions between the research community and the entities using the data collected in nanotechnology environmental, health, and safety research enhances the impact and value of NNI efforts and provides a collaborative approach.

Nanomanufacturing, Industry Liaison, and Innovation Working Group

Technology innovation is the key to realizing the NNI vision. This requires a close alliance among all the players in the innovation process, from researchers to industrial leaders, regulators, and investors. The NSET Subcommittee's Nanomanufacturing, Industry Liaison, and Innovation (NILI) Working Group coordinates many of the NNI activities in this area.

The NNI promotes U.S. leadership in the creation of new products and manufacturing processes derived from discovery at the nanoscale. The NNI creates mechanisms to facilitate nanotechnology innovation and to improve technology transfer to industry and promotes the exchange of information among Federal agencies, academia, and state, regional, and local organizations. Within the Federal Government, this effort includes interagency cooperation in the areas of standards, nomenclature, nanomanufacturing research, and use of programs that encourage innovation in small businesses.

Industry liaison groups are a key tool in this undertaking. These are partnerships between the NNI and industry sectors. These liaison groups focus on a particular industry (e.g., electronics, chemicals, forest products). Formation of an industrial liaison group is open to any industrial sector. These groups review the status of relevant R&D, and they recommend new R&D directions and activities; work to identify best technology-transfer practices; and foster development of nanotechnology standards, terminology, nomenclature, and measurement and characterization methods.

Nanotechnology Public Engagement and Communications Working Group

The NNI recognizes that the perspectives of public and stakeholder groups are vital in the nanotechnology R&D enterprise and considers public engagement to be one of its key objectives. The NSET Subcommittee's Nanotechnology Public Engagement and Communications (NPEC) Working Group, in conjunction with the NNCO, coordinates many of the NNI activities in this area.

NPEC encourages, coordinates, and supports NNI member agencies and interagency efforts toward educating and engaging the public, policy makers, and stakeholder groups regarding nanotechnology, its applications and implications, and the work of the NNI. NPEC also facilitates the identification, prioritization, and implementation of activities involving public outreach, engagement, and communications on behalf of the NSET Subcommittee by promoting responsible communication of research results and topics of special interest; environmental, health, and safety issues; ethical, legal, and social implications issues; and public engagement.

To enable better communication among the NSET Subcommittee and its working groups, individual member agencies, and the NNCO, NPEC assists in the development of information and guidance for communications activities, and supports the development of best practices for outreach and engagement among governmental and non-governmental organizations, the public, and other stakeholders regarding the responsible development of nanotechnology.

National Nanotechnology Coordination Office

The National Nanotechnology Coordination Office (NNCO) provides technical and administrative support for the NSET Subcommittee, serves as a central point of contact for Federal nanotechnology R&D activities, and provides public outreach on behalf of the NNI.

The activities of the NNCO are numerous. The NNCO coordinates the preparation and publication of NNI interagency planning, budget, and assessment documents, such as the annual NNI supplement to the President's Budget. The NNCO organizes meetings of the NSET Subcommittee and its working groups. It also organizes NNI-sponsored workshops and, as appropriate, prepares and publishes reports of those workshops. These documents, along with nanotechnology-related news, educational resources, funding opportunities, and other information, are available at the NNI website, www.nano.gov. This website also provides information about recent developments in nanotechnology and NNI activities and serves as a resource for nanotechnology-related information. The NNCO coordinates development of information on the NNI and its activities for Congress when requested. The NNCO Director arranges for the triennial review of the NNI by the National Research Council of the National Academies (NRC/NA). The NNCO will continue to organize public input and outreach activities; future examples may include web dialogues, citizens' panels, consensus conferences, and educational events.

Contributions from the NSET Subcommittee member agencies fund the NNCO. The White House Co-Chair of the NSTC Committee on Technology appoints the NNCO Director, in consultation with the Co-Chairs of the NSET Subcommittee. In accordance with the Act, the NNCO Director is detailed from a Federal agency to the NSTC as an agency representative and reports to the White House Co-Chair of the CT.

Executive Office of the President

Representatives from the Executive Office of the President (EOP) participate in NNI activities to ensure that implementation of the NNI is coordinated and consistent with

Government-wide priorities. The primary points of interaction are the Office of Science and Technology Policy (OSTP) and the Office of Management and Budget (OMB).

OSTP is responsible for advising the EOP on matters relating to science and technology and supports coordination of interagency science and technology activities. OSTP administers the NSTC, and the OSTP representative to the NSET Subcommittee is a Co-Chair of the subcommittee. This arrangement provides EOP-level input on and support for various NNI activities.

OMB is responsible for coordinating with the agency budget offices to establish the nanotechnology R&D budget for planning and tracking purposes. Each year, OMB collects budget information regarding the total Federal investment in nanotechnology R&D, as well as information about agency investments within each program component area.

ASSESSMENT

The Act calls for periodic assessment of the NNI through annual interagency reporting and review by external advisory bodies.

The Act calls for an annual interagency analysis of progress towards achieving the goals and priorities established for the NNI. This analysis is provided in the NNI supplement to the President's Budget, which also serves as the annual report called for in the Act. Specifically, progress towards achieving NNI goals and priorities is analyzed in terms of (1) coordination and cross-cutting activities, including activities of the NSET Subcommittee and (2) activities relating to the four NNI goals. Goal-related activities are: (a) NNI interagency activities, (b) NNI activities that take place within individual agencies, (c) NNI engagement with outside groups, and (d) nanotechnology-related activities external to the NNI, including international activities.

Review by outside advisory groups is vital to keeping NNI efforts focused and balanced, and the Act established two mechanism for such review. First, the Act called for the President to establish the National Nanotechnology Advisory Panel (NNAP) to advise the President and the NSTC on matters relating to the NNI. The Act specifically calls for the NNAP to assess the Federal nanotechnology R&D program at least once every two years. Executive Order 13349 designates the President's Council of Advisors on Science and Technology (PCAST) as the NNAP. The members of PCAST are senior representatives from industry and academic research institutions, with extensive experience in managing large science and technology organizations. In addition, PCAST identified a group of nanotechnology experts, which provided input and feedback on scientific and technological questions during the PCAST review. Second, the Act calls for the NRC/NA to review the NNI every third year. The NRC/NA panel for the NNI reviews is comprised of a broad cross-section of technical experts with knowledge specifically related to nanotechnology. The NRC/NA provides independent science, technology, and health policy advice to the Government. It is the principal operating agency of the National Academies in providing services to the government, the public, and the scientific and engineering communities.

The first assessment by PCAST in its role as the NNAP was released in May 2005, and the first NRC/NA review was completed in November 2006. The perspectives of these two bodies, and their assessments, are complementary, and the NNI has benefited from the diverse inputs. The resulting recommendations have led to specific actions and focused attention in areas that were highlighted by both groups, including research on health and safety aspects of nanotechnology and expanded efforts to improve education and workforce preparation. Looking forward, reviews from both PCAST and the NRC/NA are expected in 2008; the latter will focus on assessing the NNI environmental, health, and safety research strategy. The NNI member agencies will implement the recommendation of these and subsequent reviews to further strengthen the initiative.

APPENDIX A. WORKSHOPS

The following is a list of workshops sponsored or endorsed by the Nanoscale Science, Engineering, and Technology Subcommittee between 2001-2007. These workshop were organized or supported by the National Nanotechnology Coordination Office or by agencies participating in the National Nanotechnology Initiative.

Nanotechnology: Opportunities and Challenges (September 10, 2001). Report: http://www.wtec.org/nanoreports/

- Nanomanufacturing and Processing (January 5-7, 2002). Report: http://www.nsf.gov/mps/dmr/nsfec workshop report.pdf
- Nanotechnology Innovation for Chemical, Biological, Radiological, and Explosive (CBRE): Detection and Protection (May 2-3, 2002).
- Report: http://www.wtec.org/nanoreports/ From the Laboratory to New Commercial Frontiers (May 23, 2002). Report: http://www.wtec.org/nanoreports/
- Chemical Industry R&D Roadmap for Nanomaterials by Design: From Fundamentals to Function (September 30—October 1, 2002).
- Report: http://www.chemicalvision2020.org/nanomaterialsroadmap.html
- Nanoscale Science and Engineering for Agriculture and Food Systems (November 18-19, 2002).
- Report: http://www.csrees.usda.gov/nea/technology/in focus/nanotech if workshop.html
- Buildings for Advanced Technology (January 14-16, 2003). Overview: http://www.nanobuildings.com/badoverview/
- Nanotechnology Grand Challenge in the Environment (May 8-10, 2003). Report: http://www.nano.gov/html/res/pubs.html
- Materials by Design (June 11-13, 2003).
- Nanotechnology and the Environment: Applications and Implications (September 15-16, 2003). Summary: http://es.epa.gov/ncer/nano/publications/index.html
- Regional, State, and Local Initiatives in Nanotechnology (September 30—October 1, 2003). Report: http://www.nano.gov/html/res/pubs.html
- Nanobiotechnology (October 9-11, 2003). Report: http://www.nano.gov/html/res /pubs.html
- Societal Implications of Nanoscience and Nanotechnology (December 3-5, 2003). Report: http://www.nano.gov/htmllres/pubs.html

- Instrumentation and Metrology for Nanotechnology (January 27-29, 2004). Report: http://www.nano.gov/html/res/pubs.html
- Nanoelectronics, Nanophotonics, and Nanomagnetics (February 11-13, 2004).

APPENDIX B. GLOSSARY

The Act	The 21" Century Nanotechnology Research and Development Act (Public Law 108-153) agencies Departments, Agencies, Commissions, etc. of the Executive Branch of the United States Government
BIS	Bureau of Industry and Security (DOC)
CPSC	Consumer Product Safety Commission
CS	Committee on Science of the NSTC
CSREES	Cooperative State Research, Education, and Extension Service (USDA)
CT	Committee on Technology of the NSTC
DHS	Department of Homeland Security
DOC	Department of Commerce
DOD	Department of Defense
DOE	Department of Energy
DOEd	Department of Education
DOI	Department of the Interior
DOJ	Department of Justice
DOL	Department of Labor
DOS	Department of State
DOT	Department of Transportation
DOTreas	Department of the Treasury
EHS	Environmental, health, and safety
ELSI	Ethical, legal, and social implications
EOP	Executive Office of the President
EPA	Environmental Protection Agency
FDA	Food and Drug Administration (HHS)
FS	Forest Service (USDA)
GIN	Global Issues in Nanotechnology Working Group of the NSET Subcommittee
HHS	Department of Health and Human Services
IARPA	Intelligence Advanced Research Projects Activity
ITC	International Trade Commission
NASA	National Aeronautics and Space Administration
NCI	National Cancer Institute (HHS/NIH)
NCL	Nanotechnology Characterization Laboratory (HHS/NIH/NCI)
NEHI	Nanotechnology Environmental and Health Implications Working Group of the NSET Subcommittee
NIEHS	National Institute of Environmental Health Sciences (HHS/NIH)
NIH	National Institutes of Health (HHS)

NILI	Nanomanufacturing, Industry Liaison, and Innovation Working Group of the NSET Subcommittee
NIOSH	National Institute for Occupational Safety and Health (HHS/Centers for Disease Control and Prevention)
NIST	National Institute of Standards and Technology (DOC)
NNAP	National Nanotechnology Advisory Panel
NNCO	National Nanotechnology Coordination Office
NNI	National Nanotechnology Initiative
NPEC	Nanotechnology Public Engagement and Communications Working Group of the NSET Subcommittee
NRC	Nuclear Regulatory Commission
NRC/NA	National Research Council of the National Academies
NSET	Nanoscale Science, Engineering, and Technology Subcommittee of the NSTC
NSF	National Science Foundation
NSRC	Nanoscale Science Research Center (funded by DOE)
NSTC	National Science and Technology Council
OMB	Office of Management and Budget (EOP)
OSTP	Office of Science and Technology Policy (EOP)
PCA	Program Component Area
PCAST	President's Council of Advisors on Science and Technology
USDA	U.S. Department of Agriculture
USGS	U.S. Geological Survey (DOI)
USPTO	U.S. Patent and Trademark Office (DOC)

In: Nanotechnology in the USA: Developments, Policies and Issues ISBN 978-1-60692-800-4
Editor: Carl H. Jennings

Chapter 2

THE NATIONAL NANOTECHNOLOGY INITIATIVE: OVERVIEW, REAUTHORIZATION, AND APPROPRIATIONS ISSUES*

John F. Sargent

ABSTRACT

Nanotechnology — a term encompassing the science, engineering, and applications of submicron materials — involves the harnessing of unique physical, chemical, and biological properties of nanoscale substances in fundamentally new and useful ways. The economic and societal promise of nanotechnology has led to substantial and sustained investments by governments and companies around the world. In 2000, the United States launched the world's first national nanotechnology program. Since then, the federal government has invested more than $8 billion in nanoscale science, engineering, and technology through the U.S. National Nanotechnology Initiative (NNI). U.S. companies and state governments have invested billions more. As a result of this focus and these investments, the United States has, in the view of many experts, emerged as a global leader in nanotechnology. However, the competition for global leadership in nanotechnology is intensifying as countries and companies around the world increase their investments.

Nanotechnology' s complexity and intricacies, early stage of development (with commercial pay-off possibly years away), and broad scope of potential applications engender a wide range of public policy issues. Maintaining U.S. technological and commercial leadership in nanotechnology poses a variety of technical and policy challenges, including development of technologies that will enable commercial scale manufacturing of nanotechnology materials and products; environmental, health, and safety (EHS) concerns; and maintenance of public confidence in its safety.

Congress established programs, assigned responsibilities, and initiated research and development (R&D) related to these issues in the 21st Century Nanotechnology Research and Development Act of 2003 (P.L. 108-153). While many provisions of this act have no sunset provision, FY2008 is the last year of agency authorizations included in the act. Consideration may be given to reauthorization of this act in 2008.

---

* This is an edited, excerpted and augmented edition of a CRS Report RL34401,dated February 29, 2008.

Proponents of the NNI assert that nanotechnology is one of the most important emerging and enabling technologies and that U.S. competitiveness, technological leadership, national security, and societal interests require an aggressive approach to the development and commercialization of nanotechnology.

Critics of the NNI voice concerns that reflect disparate underlying beliefs. Some critics assert that the government is not doing enough to move technology from the laboratory into the marketplace. Others argue that the magnitude of the public investment may skew what should be market-based decisions in research, development, and commercialization. Still other critics say that the inherent risks of nanotechnology are not being addressed in a timely or effective manner.

From the NNI's inception through FY2008, Congress has appropriated a total of $8.5 billion for NNI activities. NNI funding in FY2008 is estimated to be $1.491 billion. For FY2009, President Bush requested $1.527 billion for the NNI.

Introduction

Nanotechnology has been an issue of interest to Congress for a number of years, coming into focus in 2000 with the launch of the U.S. National Nanotechnology Initiative (NNI) by President Clinton in his FY2001 budget request to Congress. Since then, Congress has appropriated more than $8 billion for nanotechnology research and development (R&D). These efforts have been directed at advancing understanding and control of matter at the nanoscale,[1] where the physical, chemical, and biological properties of materials differ in fundamental and useful ways from the properties of individual atoms or bulk matter.[2]

The development and application of nanotechnology — more fully explained below — across a wide array of products and industries holds the potential for significant economic and societal benefits. To capture these benefits, the United States will have to effectively address a variety of technical and policy challenges that stand as potential barriers to commercialization, including environmental, health, and safety concerns and their implications for workplace, environmental, food, and drug regulations; development of standards, reference materials, and consistent nomenclature; development of new measurement methods and tools; effective technology transfer to the private sector; protection of intellectual property; availability, affordability, and patience of investment capital; ethical, legal, and societal concerns; public understanding, support, and acceptance; and development of a world-class scientific and technical nanotechnology workforce.

In 2003, Congress passed the 21st Century Nanotechnology Research and Development Act (P.L. 108-153) providing a legislative foundation for some of the activities of the NNI, authorizing agency funding levels through FY2008, and intended to address several of these challenges. Discussions are underway with respect to the possible reauthorization of this act in 2008. Congress may use this opportunity to further address these issues and to establish authorization levels for agency nanotechnology R&D. Alternatively, Congress may choose to address these issues in separate legislation. Several bills were introduced in the first session of the 1 10th Congress to address specific nanotechnology issues.

This chapter provides an overview of nanotechnology, the National Nanotechnology Initiative, possible reauthorization of the 21st Century Nanotechnology Research and Development Act of 2003 (P.L. 108-153), and appropriations issues

OVERVIEW

Nanotechnology: A Description

The term "nanotechnology" is often used as an all-encompassing term for nanoscale science, engineering, and technology. Nanotechnology is the understanding and control of matter at dimensions of roughly 1 to 100 nanometers, the size-scale between individual atoms and bulk materials, where unique phenomena enable novel applications. A nanometer is one-billionth of a meter, or about the width of 10 hydrogen atoms arranged side-by-side in a line. Nanotechnology involves imaging, measuring, modeling, and manipulating matter at this size-scale.

At the nanoscale, the physical, chemical, and biological properties of materials differ in fundamental and useful ways from the properties of individual atoms and molecules or bulk matter. Nanotechnology R&D is directed toward understanding and creating improved materials, devices, and systems that exploit these new properties.

Physicist Richard Feynman's remarks at the 1959 annual meeting of the American Physical Society are often cited as the first articulation of and vision for nanotechnology. Though he did not use the term nanotechnology in this speech, he spoke of controlling matter at the nanoscale and creating atomic-level machines, positing some of the applications that doing so might enable.

Source: The National Nanotechnology Initiative Strategic Plan, 2004, Nanoscale Science, Engineering, and Technology Subcommittee, National Science and Technology Council, The White House, December 2004.

The economic and societal promise of nanotechnology has led to involvement and investments by governments and companies around the world. In 2000, the United States became the first nation to establish a formal, national initiative to advance nanoscale science, engineering, and technology — the National Nanotechnology Initiative. Since then, Congress has appropriated more than $8 billion in nanoscale science, engineering, and technology through the NNI. U.S. companies and state governments have invested billions more.

As a result of this focus and these investments, the United States has emerged as a global leader in nanotechnology. However, the competition for global leadership is intensifying as foreign investments in nanoscale science, engineering, and technology increase. Other nations have followed the U.S. lead and established their own national nanotechnology programs, each with varying degrees of investment, foci, and support for industrial applications and commercialization. Today, almost every nation that supports R&D has a national-level nanotechnology program.

Global nanotechnology investment in 2006 has been estimated at $12.4 billion, with public investments accounting for approximately $6.4 billion.[3] While the United States leads all other nations in public investments in nanotechnology R&D, it is estimated to account for only about a quarter of global annual public investments.[4]

Global investments in nanotechnology already have begun to yield economic benefits as products incorporating nanotechnology enter the marketplace. These products are estimated to have produced $50 billion in revenues in 2006.[5] By tapping the unique properties that emerge at the nanoscale, proponents maintain that nanotechnology holds the potential for products that could transform existing industries and create new ones, clean and protect the environment, extend and improve the quality of our lives, and strengthen the national security. Most nanotechnology products currently on the market — such as faster computer processors, higher density memory devices, lighter-weight auto parts, stain-resistant clothing, antibiotic bandages, cosmetics, and clear sunscreen — are evolutionary in nature, offering incremental improvements in characteristics such as performance, aesthetics, cost, size, and weight.

Evolutionary nanotechnology products, however, represent only a small fraction of what many see as the substantial longer-term economic and societal promise of nanotechnology. One estimate projects nanotechnology product revenues will reach $2.6 trillion6 by 2014, or 15% of global manufacturing output, while another estimates global revenues will reach $2.95 trillion by 2015, of which almost half will come from semiconductors.[7, 8]

Many nanotechnology advocates — including business executives, scientists, engineers, medical professionals, and venture capitalists — assert that in the longer term, nanotechnology, especially in combination with information technology, biotechnology, and the cognitive sciences, may deliver revolutionary advances, including:

- new prevention, detection, and treatment technologies that could reduce substantially death and suffering from cancer and other deadly illnesses;[9]
- new organs to replace damaged or diseased ones;[10]
- contact lenses, skin patches, and glucose-sensing tattoos that monitor diabetics' blood sugar levels and warn when too high or low;[11]
- clothing that protects against toxins and pathogens;[12]
- clean, inexpensive, renewable power through energy creation, storage, and transmission technologies;[13]
- inexpensive, portable water purification systems that provide universal access to safe water;[14]
- energy efficient, low-emission "green" manufacturing systems;[15]
- high-density memory systems capable of storing the entire Library of Congress collection on a device the size of a sugar cube;[16]
- agricultural technologies that increase crop yield and improve nutritional value, reducing global hunger and malnutrition;[17]
- self-repairing materials;[18]
- powerful, small, inexpensive sensors that can warn of minute levels of toxins and pathogens in air, soil, or water;[19] and
- decontaminated industrial sites through environmental remediation.[20]

Although some applications of nanotechnology have proven market-ready, much fundamental research remains ahead, including efforts to advance understanding of nanoscale phenomena; characterize nanoscale materials; understand how to control and manipulate nanoscale particles; develop instrumentation and measurement methods; and understand how nanoscale particles interact with humans, animals, plants, and the environment. In addition, several federal agencies — such as the Departments of Defense, Energy, and Homeland Security — see the potential for nanotechnology to help address mission requirements.

Historically, the federal government has played a central role in funding these types of research and development activities.

Though federal nanoscale science, engineering, and technology R&D had been underway for over a decade, the NNI was first initiated as a Presidential technology initiative in 2000.[21] The original participating agencies were the National Science Foundation (NSF), the Department of Defense (DOD), the Department of Energy (DOE), the Department of Commerce's (DOC) National Institute of Standards and Technology (NIST), the National Aeronautics and Space Administration (NASA), and the Department of Health and Human Services' National Institutes of Health (NIH). In 2007, 25 agencies participated in the NNI, including 13 that received appropriations to conduct and/or fund nanotechnology R&D.

Since its first year of funding in FY2001, the NNI's annual appropriations have grown three-fold to an estimated $1.491 billion in FY2008. From FY2001 through FY2008, Congress appropriated a total of $8.5 billion for NNI activities. President Bush has requested $1.527 billion for the NNI in FY2009.

In 2003, Congress provided a statutory foundation for some of the activities of the NNI through the 21st Century Nanotechnology Research and Development Act of 2003 (P.L. 108-153). The act established a National Nanotechnology Program (NNP) and provided authorizations for a subset of the NNI agencies, namely the NSF, DOE, NASA, NIST, and Environmental Protection Agency (EPA). The act, however, did not address the participation of several agencies that fund nanotechnology R&D under the NNI, including DOD, NIH, and the Department of Homeland Security (DHS). Nevertheless, coordination of nanotechnology R&D activities across all NNI funding agencies continues under the National Science and Technology Council's (NSTC's) Nanoscale Science, Engineering, and Technology (NSET) subcommittee.[22] According to the NSET subcommittee's 2004 NNI Strategic Plan, "For continuity and to capture this broader participation, the coordinated federal activities as a whole will continue to be referred to as the National Nanotechnology Initiative." Accordingly, the functions and activities established under the act are incorporated into the Executive Branch's implementation of the NNI.

While many provisions of this act have no sunset provision, FY2008 is the last year of agency authorizations included in the act. Discussions are underway regarding the potential reauthorization of this act in 2008.

The thrust of the NNI has primarily been the development of fundamental scientific knowledge through basic research. Investments at mission agencies, such as DOD, have supported nanotechnology applications development for which they are a primary customer. Other investments have supported infrastructural technologies. For example, NIST has contributed to developing tools and standards that enable measurement and control of matter at the nanoscale, thereby supporting the conduct of R&D and the ability to manufacture nanoscale materials and products. As understanding of nanotechnology has matured, the NNI has worked with a variety of industry organizations to facilitate the movement of research results from the laboratory bench to the marketplace in fields as disparate as semiconductors, chemicals, energy, concrete, and forest products.

The NNI agencies also have begun to address research needs and regulatory issues related to environmental, health, and safety issues, as well as issues such as public understanding and workforce education and training. The NNI agencies actively engage in a variety of international fora, such as the Organization for Economic Cooperation and Development (OECD) and the International Standards Organization (ISO), to cooperatively address nanotechnology issues related to EHS, metrology [23] and standards, nomenclature, and nanoscale materials characterization.

Maintaining U.S. leadership poses a variety of technical, economic, and policy challenges, including:

- safeguarding the environment and ensuring human health and safety;
- creating the standards, reference materials, nomenclature, methods, and tools for metrology to enable the manufacturing of nanoscale materials and products;
- developing a world-class scientific and technical nanotechnology workforce;
- translating research results into products, including effective technology transfer to the private sector;
- understanding public perceptions and attitudes and fostering public understanding;
- addressing ethical, legal and societal implications;
- protecting intellectual property;
- securing investment capital for early-stage research, development, and commercialization; and
- fostering and facilitating international cooperation and coordination.

Proponents of the NNI assert that nanotechnology is one of the most important emerging and enabling technologies [24] and that U.S. competitiveness, technological leadership, national security and societal interests require an aggressive approach to the development and commercialization of nanotechnology. Critics of the NNI hold a variety of competing views, asserting that government is not doing enough, is doing too much, or is moving too quickly.

Some in industry have criticized the NNI for being overly focused on basic research and not being aggressive enough in moving NNI-funded R&D out of government and university laboratories and into industry. Others in industry have criticized the federal government for not providing mechanisms to help advance nanotechnology R&D to the point where it becomes economically viable for venture capitalists, corporations, and other investors to create products and bring them to market. Some refer to this gap as the "valley of death."[25] Still others in industry have criticized the NNI for not adequately supporting the development of metrology, standards, equipment, and processes necessary to manufacture nanotechnology materials, products, and systems at a commercial scale.

Conversely, supporters of industry-driven market investments contend that extensive government support for nanotechnology may supplant the judgment of the marketplace by picking "winners and losers" in technological development. For example, the size and directions of the NNI investments may encourage industry to follow the government's lead rather than independently selecting R&D directions itself or, alternatively, may result in the promotion of a less effective technology path over a more effective one. These supporters also assert that federal government funding of scientific research is often wasteful, driven by political considerations and not scientific merit.[26]

Some non-governmental organizations (NGO) are critical of nanotechnology for its potential adverse impacts on human health and safety and on the environment. They assert that the government is pushing ahead too quickly in developing nanotechnology and encouraging its commercialization and use without adequately investing in research focused on understanding and mitigating negative EHS implications.[27] They argue that the very characteristics that make nanotechnology promising also present significant potential risks to human health and safety and the environment. Some of these critics argue for application of the "precautionary principle," which holds that regulatory action may be required to control potentially hazardous substances even before a causal link has been established by scientific evidence.[28] At least one NGO has called for a moratorium on nanotechnology R&D and new commercial products incorporating synthetic nanoparticles.[29]

National Nanotechnology Initiative

The National Nanotechnology Initiative is an interagency program that coordinates federal nanoscale science, engineering, and technology R&D activities and related efforts among participating agencies.

Vision and Goals

The National Science and Technology Council (NSTC) has stated the following vision for the NNI:

> A future in which the ability to understand and control matter on the nanoscale leads to a revolution in technology and industry. The NNI will expedite the discovery, development, and deployment of nanotechnology in order to achieve responsible and sustainable economic benefits, to enhance the quality of life, and to promote national security.[30]

To achieve its vision, the NNI has established four goals: maintain a world-class R&D program aimed at realizing the full potential of nanotechnology; facilitate transfer of new technologies into products that provide economic growth, jobs, and other public benefits; develop educational resources, a skilled workforce, and the supporting infrastructure and tools to advance nanotechnology; and support responsible development of nanotechnology.[31]

History

Attempts to coordinate federal nanoscale R&D began in November 1996, as staff members from several agencies met regularly to discuss their plans and programs in nanoscale science and technology. This group continued informally until September 1998, when it was designated as the Interagency Working Group on Nanotechnology (IWGN) under the NSTC. In August 1999, IWGN completed its first draft of a plan for an initiative in nanoscale science and technology, which was subsequently approved by the President's Council of Advisors on Science and Technology (PCAST) and the White House Office of Science and Technology Policy (OSTP).[32]

In his 2001 budget submission to Congress, then-President Clinton raised nanotechnology-related research to the level of a federal initiative, officially referring to it as the National Nanotechnology Initiative.[33]

Legislative Approach

Congress has played a central role in the National Nanotechnology Initiative, providing appropriations for the conduct of nanoscale science, engineering, and technology research; establishing programs; and creating a legislative foundation for the activities of the NNI.

Congressional funding for the NNI is provided through appropriations to each of the NNI-participating agencies. The NNI has no centralized funding. The overall NNI budget is calculated by aggregating the nanotechnology budgets for each of the federal agencies that conduct or provide funding for nanoscale science, engineering, and technology research.

In FY2001, the first year of NNI funding, Congress provided $464 million to eight agencies for nanoscale science, engineering, and technology research.[34] The NNI has continued to receive support from both Congress and the White House. Both the number of agencies participating in the NNI and the size of the federal investment have grown. Today 25 agencies participate in the NNI, 13 of which received appropriated funds for nanotechnology R&D in FY2007.[35] Total NNI funding in FY2008 was $1.49 1 billion, more than three times the level of funding provided in FY2001. The original six agencies identified at the launch of the NNI [36] still account for the vast majority of NNI funding, 98.3% in FY2008.

Some of the NNI's activities were codified and further defined in the 21st Century Nanotechnology Research and Development Act of 2003 which was passed by Congress in November 2003. On December 3, 2003, the act was signed into law (P.L. 108-153) by President Bush.[37] The legislation received strong bipartisan support in both the House of Representatives, which passed the bill on a recorded vote of 405-19, and in the Senate, which passed the bill by unanimous consent.

Though this act is often referred to as the enabling legislation for the National Nanotechnology Initiative, the act actually establishes a National Nanotechnology Program (NNP). The act provides authorizations for five NNI agencies — the National Science Foundation, Department of Energy, NASA, National Institute of Standards and Technology, and Environmental Protection Agency — but not for the Department of Defense, National Institutes of Health, Department of Homeland Security,[38] or other NNI research agencies that collectively accounted for 46% of NNI funding in FY2003.

The act created the NNP for the purposes of establishing the goals, priorities, and metrics for evaluation of federal nanotechnology research, development, and other activities; investing in federal R&D programs in nanotechnology and related sciences to achieve those goals; and providing for interagency coordination of federal nanotechnology research, development, and other activities undertaken pursuant to the NNP.

Key provisions of the act include:

- authorizing appropriations for the nanotechnology-related activities of the National Science Foundation, Department of Energy, NASA, National Institute of Standards and Technology, and Environmental Protection Agency for fiscal years 2005 through 2008, totaling $3.679 billion for the four year period;

- establishing a National Nanotechnology Coordination Office, with a director and full time staff to provide administrative support to the NSTC;
- establishing a National Nanotechnology Advisory Panel (NNAP) to advise the President and the NSTC on matters relating to the NNP.
- establishing a triennial review of the NNP by the National Research Council of the National Academies of Sciences;
- directing the NSTC to oversee the planning, management, and coordination of the program, including the development of a triennial strategic plan;
- directing the Department of Commerce's National Institute of Standards and Technology to establish a program to conduct basic research on issues related to the development and manufacture of nanotechnology, and to use the Manufacturing Extension Partnership program to ensure results reach small- and medium-sized manufacturing companies;
- directing the Secretary of Commerce to use the National Technical Information Service to establish a clearinghouse of information related to commercialization of nanotechnology research;
- directing the Secretary of Energy to establish a program to support consortia to conduct interdisciplinary nanotechnology R&D designed to integrate newly developed nanotechnology and microfluidic tools with systems biology and molecular imaging;
- directing the Secretary of Energy to carry out projects to develop, plan, construct, acquire, operate, or support special equipment, instrumentation, or facilities for investigators conducting nanotechnology R&D; and
- directing the establishment of two centers, on a merit-reviewed and competitive basis: (1) the American Nanotechnology Preparedness Center, to conduct, coordinate, collect, and disseminate studies on the societal, ethical, environmental, educational, legal, and workforce implications of nanotechnology; and to identify anticipated issues related to the responsible research, development, and application of nanotechnology, as well as provide recommendations for preventing or addressing such issues, and (2) the Center for Nanomaterials Manufacturing, to encourage, conduct, coordinate, commission, collect, and disseminate research on new manufacturing technologies for materials, devices, and systems with new combinations of characteristics, such as, but not limited to, strength, toughness, density, conductivity, flame resistance, and membrane separation characteristics; and to develop mechanisms to transfer such manufacturing technologies to U.S. industries.

While the act establishes a National Nanotechnology Program, the Executive Branch continues its broader effort under the NNI framework and name. According to the NNI's 2004 Strategic Plan:

Many of the activities outlined in the Act were already in progress as part of the NNI. Moreover, the ongoing management of the initiative involves considerable input from Federal agencies that are not named specifically in the Act.... For continuity, and to capture this broader participation, the coordinated Federal activities as a whole will continue to be referred to as the National Nanotechnology Initiative.[39]

STRUCTURE

Nanoscale Science, Engineering, and Technology Subcommittee

The NNI is coordinated within the White House through the NSTC, the Cabinet-level council by which the President coordinates science, space, and technology policies across the federal government. Operationally, NNI coordination is accomplished through the Nanoscale Science, Engineering, and Technology (NSET) subcommittee of the NSTC's Committee on Technology (CT). The NSET subcommittee also has an informal reporting relationship to the NSTC's Committee on Science (CS). The NSET subcommittee is led by an agency co-chair, currently from the Department of Energy (DOE), and an OSTP co-chair. The NSET subcommittee is comprised of representatives from 25 federal entities, OSTP and the Office of Management and Budget.[40]

The NSET subcommittee has established several chartered and non-chartered working groups that conduct work in key subject areas. The three chartered working groups are:[41]

National Environmental and Health Implications (NEHI)

The NEHI working group was chartered to provide for exchange of information among agencies that support research and those responsible for regulations and guidelines related to nanotechnology products; to facilitate identification, prioritization, and implementation of research and other activities required for the responsible research, development, utilization, and oversight of nanotechnology; and to promote communication of information related to research on environmental and health implications of nanotechnology to other government agencies and non-government parties. To this end, the NEHI working group has been attempting to identify and prioritize environmental, health, and safety research needs related to nanotechnology. Twenty of the 25 NNI agencies participate in the NEHI working group, and 13 agencies fund safety-related nanotechnology research and/or have regulatory authorities to guide the safe use of nanomaterials.[42]

National Innovation and Liaison with Industry (NILI)

The NILI working group was chartered to enhance collaboration and information sharing between U.S. industry and government on nanotechnology-related activities. It also facilitates federal, regional, state, and local nanotechnology R&D and commercialization activities. In addition, the NILI working group is to create innovative methods for transferring federally funded technology to industry. The NILI working group has facilitated collaborations between the NNI and the semiconductor/electronics industry, chemical industry, forest products industry, and the Industrial Research Institute.[43]

Global Issues in Nanotechnology (GIN)

The GIN working group was chartered to monitor foreign nanotechnology programs and development; broaden international collaboration on nanotechnology R&D, including safeguarding the environment and human health; and promote U.S. commercial and trade interests in nanotechnology. The NEHI working group works with the GIN working group to coordinate the U.S. position and participation in international activities related to environmental, health, and safety implications of nanotechnology. The GIN working group facilitates international collaboration on pre-competitive and non-competitive aspects of nanotechnology, and international engagement on trade, commercialization and regulatory issues.

In addition to the chartered working groups, the NSET subcommittee has two non-chartered working groups: the Nanomanufacturing working group and the Nanotechnology Public Engagement and Communications (NPEC) working group.

National Nanotechnology Coordination Office

The National Nanotechnology Coordination Office (NNCO) provides administrative and technical support to the NSET subcommittee. Initially established in 2001 through a memorandum of understanding among the NNI participating agencies,[44] the NNCO was authorized by the 21st Century Nanotechnology Research and Development Act of 2003 (P.L. 108-153). The NNCO was charged under the act with providing technical and administrative support to the NSTC and NNAP; serving as the point of contact for information on Federal nanotechnology activities for the exchange of technical and programmatic information among stakeholders; conducting public outreach; and promoting access to and early application of NNP technologies, innovation, and expertise.

The act authorizes the work of the NNCO to be funded by contributions from NSET subcommittee member agencies. According to the NNCO, funding is provided through a memorandum of understanding signed by eight NNI agencies.[45] In principle, each agency contributes to the NNCO budget in proportion to their share of the President's total nanotechnology budget request for the signatory agencies. However, two of the signatories, EPA and DOT, had sufficiently small enough nanotechnology budgets in the early years of the NNI that they were not expected to contribute. EPA now contributes to funding the NNCO. Total NNCO funding from the agencies in FY2008 is $2.1 million.

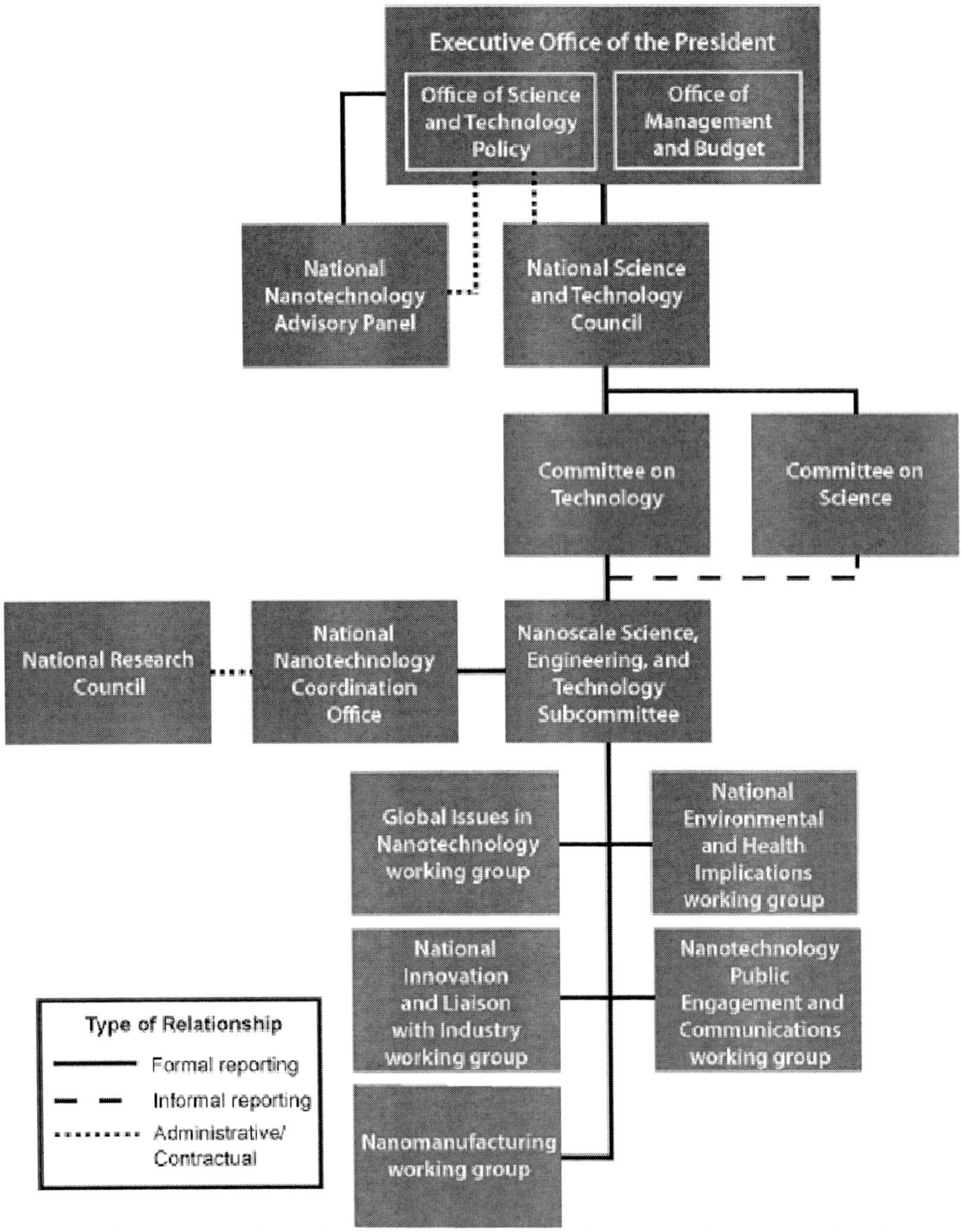

Source: The National Nanotechnology Strategic Plan, Nanoscale, Science, Engineering, and Technology Subcommittee, National Science and Technology Council, The White House, December 2007. (Nanomanufacturing working group added to chart in cited source.)

Figure 1. Organizations With a Role in the National Nanotechnology Initiative and Their Relationships.

FUNDING

The NNI supports fundamental and applied research on nanotechnology by funding research, creating multidisciplinary centers of excellence, and developing key research infrastructure. It also supports activities aimed at addressing the societal implications of nanotechnology, including ethical, legal, human and environmental health, and workforce issues.

This section provides information on NNI funding from two perspectives: organizationally by agency and functionally by program component area.

Agency Funding

The NNI budget is an aggregation of the nanotechnology components of the individual budgets of NNI-participating agencies. The NNI budget is not a single, centralized source of funds that is allocated to individual agencies. In fact, agency nanotechnology budgets are developed internally as part of each agency's overall budget development process. These budgets are subjected to review, revision, and approval by the Office of Management and Budget and become part of the President's annual budget submission to Congress. The NNI budget is then calculated by aggregating the nanotechnology components of the appropriations provided by Congress to each federal agency.

For FY2008, the NNI budget totaled an estimated $1.491 billion, a 4.8% increase over FY2007 funding and more than triple the $464 million federal investment in nanotechnology research in 2001. This growth in nanotechnology R&D investments reflects expectations in Congress and in the executive branch that the NNI will expand fundamental knowledge and make important contributions to national priorities. In his FY2009 budget, the President has requested $1 .527 billion for nanotechnology R&D, a 2.3% increase above the estimated FY2008 funding level. The chronology of NNI funding is detailed in Table 1.

The President's proposed FY2009 NNI budget supports a broad range of programs among 13 agencies. Agencies with the largest budgets are:

- NSF, which supports fundamental nanotechnology research across science and engineering disciplines;
- DOD, whose investments in nanotechnology are aimed at addressing the department's national security mission;
- DOE, which supports nanotechnology research providing a basis for new and improved energy efficiency, production, storage, and transmission technologies;
- NIH, which emphasizes nanotechnology-based biomedical advances occurring at the intersection of biology and the physical sciences; and
- NIST, which focuses on research in instrumentation, measurement, standards, characterization, and nanomanufacturing.

Other agencies investing in mission-related nanotechnology R&D are NASA, EPA, the Cooperative State Research, Education, and Extension Service (CSREES) and Forest Service at the Department of Agriculture (USDA), National Institute of Occupational Safety and

Health (NIOSH), DHS, Department of Justice (DOJ), and Department of Transportation's (DOT's) Federal Highway Administration (FHWA).

Table 1. NNI Funding, by Agency: FY2001-FY2008 and FY2009 Request (in millions of current dollars)

Agency	FY 2001 Actual	FY 2002 Actual	FY 2003 Actual	FY 2004 Actual	FY 2005 Actual	FY 2006 Actual	FY 2007 Actual	FY 2008 Estimate	FY 2009 Request
NSF	150	204	221	256	335	360	389	389	397
DOD[a]	125	224	322	291	352	424	450	487	431
DOE	88	89	134	202	208	231	236	251	311
NIH (DHHS)	40	59	78	106	165	192	215	226	226
NIST (DOC)	33	77	64	77	79	78	88	89	110
NASA	22	35	36	47	45	50	20	18	19
EPA	5	6	5	5	7	5	8	10	15
DOJ	1	1	1	2	2	<1	2	2	2
DHS		2	1	1	1	2	2	1	1
CSREES (USDA)			1	2	3	4	4	6	3
NIOSH					3	4	7	6	6
Forest Service (USDA)						2	3	5	5
FHWA (DOT)						1	1	1	1
TOTAL[b]	464	697	863	989	1,200	1,351	1,425	1,491	1,527

Sources: The National Nanotechnology Initiative: Research and Development Leading to a Revolution in Technology and Industry, Supplement to the President's FY2008 Budget, Nanoscale Science, Engineering, and Technology Subcommittee, National Science and Technology Council, The White House, July 2007. The National Nanotechnology Initiative Strategic Plan, Nanoscale Science, Engineering, and Technology Subcommittee, National Science and Technology Council, The White House, December 2004. National Nanotechnology Initiative: FY2009 Budget & Highlights, National Science and Technology Council, The White House, February 2008.

a. According to NSTC, the Department of Defense budgets shown above for FY2006 and FY2007 include Congressionally directed funding of approximately $76 million and $63 million, respectively. According to NSTC, the 2008 DOD estimate "includes many earmarks that are outside the NNI plan."

b. Numbers may not add due to rounding of agency budget figures.

Program Component Area Funding

The 21st Century Nanotechnology R&D Act of 2003 called for the NSET subcommittee to develop categories of investment called Program Component Areas (PCA) to provide a means by which Congress and the executive branch can be informed of and direct the relative investments in these areas. The PCAs are categories of investments that cut across the needs and interests of individual agencies and contribute to the achievement of one or more of the NNI's goals. The 2004 NNI strategic plan identified seven PCAs. The 2007 NNI strategic plan splits the seventh PCA, Societal Dimensions, into two PCAs: Environment, Health, and Safety; and Education and Societal Dimensions. A description of the seven initial PCAs and their current funding are provided below,[46] as well as a description of the two derivative PCAs.[47] The chronology of NNI funding by PCA is detailed in Table 2.

Table 2. NNI Funding, by Program Component Area, FY2006-FY2009 (in millions of current dollars)

PCA	FY2006 Actual	FY2007 Actual	FY2008 Estimate	FY2009 Request
Fundamental Phenomena and Processes	455.9	480.6	531.6	550.8
Nanomaterials	265.1	258.3	254.7	227.2
Nanoscale Devices and Systems	319.6	344.7	342.3	327.0
Instrumentation Research, Metrology, and Standards	51.0	52.5	60.4	81.5
Nanomanufacturing	33.8	48.1	50.2	62.1
Major Research Facilities and Instrumentation Acquisition	152.4	152.4	154.4	161.3
Societal Dimensions	73.5			
- Environment, Health, and Safety		48.3	58.6	76.4
- Education and Societal Dimensions		39.2	39.0	40.7
TOTAL	1,351.2	1,424.1	1,491.2	1,527.0

Source: The National Nanotechnology Initiative: Research and Development Leading to a Revolution in Technology and Industry, Supplement to the President's FY2008 Budget, Nanoscale Science, Engineering, and Technology Subcommittee, National Science and Technology Council, The White House, July 2007.

Fundamental Phenomena and Processes

Fundamental Phenomena and Processes includes investments in the discovery and development of fundamental knowledge pertaining to the new phenomena in the physical, biological, and engineering sciences that occur at the nanoscale, as well as in understanding and articulation of scientific and engineering principles related to nanoscale structures, processes, and mechanisms.

FY2008 funding for Fundamental Phenomena and Processes rose to $531.6 million, up $51.0 million (10.6%) over the FY2007 level due to increases in DOD (up $48.6 million, 23.1%) and NIH (up $9.9 million, 21.7%) funding in this PCA. The increases in FY2008 funding for DOD and NIH in this PCA were partially offset by decreases in other agencies' budgets. The President's FY2009 budget proposes $550.8 million in funding for this PCA, up $19.2 million (3.6%) above the FY2008 level.[48]

Nanomaterials. Nanomaterials includes research investments to discover novel nanoscale and nanostructured materials. This PCA also attempts to understand the properties of nanomaterials, and supports R&D to enable the design and synthesis, in a controlled manner, of nanoscale materials with targeted properties.

FY2008 funding for Nanomaterials fell to $254.7 million, down $3.6 million (1.4%) from the FY2007 level, led by a decrease in DOD funding in this PCA (down $17.1 million, 19.9%). The decline in FY2008 spending in this PCA resulting from the DOD reduction was partially offset by increases in other agencies' budgets, including a $9.0 million (13.1%) increase in DOE funding and a $3.7 million (6.3%) increase in NSF funding. The President's FY2009 budget proposes $227.2 million for this PCA, a decrease of $27.5 million (10.8%) from the FY2008 level.[49]

Nanoscale Devices and Systems. Nanoscale Devices and Systems include R&D investments that apply nanoscale science and engineering principles to create novel devices and systems or to improve existing ones. It also includes the use of nanoscale or nanostructured materials to achieve improved performance or new functionality. To meet this definition, the enabling science and technology must be at the nanoscale, but the systems and devices are not restricted to that size.

Funding for Nanoscale Devices and Systems fell to $342.3 million in FY2008, down $2.4 million (0.7%) from the FY2007 level. The President's FY2009 budget proposes $327.0 million in funding for this PCA, a decrease of $15.3 million (4.5%) from the FY2008 level, largely due to reductions in DOD (down $12.1 million, 10.1%) and DOE (down $4.9 million, 37.7%) funding in this PCA. The decrease in DOD and DOE funding for this PCA is somewhat offset by increases in other agencies' budgets.[50]

Instrumentation Research, Metrology, and Standards

The Instrumentation Research, Metrology, and Standards PCA includes R&D investments for development of tools needed to advance nanotechnology research and commercialization. Instrumentation for characterization, measurement, synthesis, and design of nanotechnology materials, structures, devices, and systems is funded through this PCA. R&D and other activities related to development of standards, including standards for nomenclature, materials, characterization, testing, and manufacture are also in this PCA.

FY2008 funding for Instrumentation Research, Metrology, and Standards increased to $60.4 million, up $7.9 million (15.0%) over the FY2007 level, led by a $3.7 million (86.0%) increase in DOD funding. The President's FY2009 budget proposes $81.5 million in funding for this PCA, an increase of $21.1 million (34.9%) above the FY2008 level, with DOE and NIST accounting for the largest increases.[51]

Nanomanufacturing. Nanomanufacturing R&D supports the development of scalable, reliable, cost-effective manufacturing of nanoscale materials, structures, devices, and systems. It also includes R&D and integration of ultra-miniaturized top-down processes and complex bottom-up processes.[52]

FY2008 funding for Nanomanufacturing rose to $50.2 million, up $2.1 million (4.4%) over the FY2007 level due to increases in the budgets of NIST and DOE. The President's FY2009 budget proposes $62.1 million for this PCA, an increase in funding of $1 1.9 million (23.7%) above the FY2008 level, due to increases in DOD, DOE, and NIST funding for this PCA.[53]

Major Research Facilities and Instrumentation Acquisition. This PCA includes investments in the establishment and ongoing operations of user facilities and networks, the acquisition of major instrumentation, and other activities related to infrastructure for the conduct of nanoscale science, engineering, and technology R&D.

FY2008 funding for Major Research Facilities and Instrumentation Acquisition rose to $154.4 million, an increase of $2.0 million (1.3%) over the FY2007 level. The President's FY2009 budget proposes $161.3 million for this PCA, an increase of $6.9 million (4.5%) above the FY2008 level, led by a $9.2 million (10.0%) increase in DOE funding and partially offset by decreases in other agencies' budgets.[54]

Societal Dimensions. The Societal Dimensions PCA includes investments in research and other activities that address the broad implications of nanotechnology to society. This includes assessing benefits and risks through research directed at environmental, health, and safety impacts of nanotechnology development; risk assessment of such impacts; education-related activities, such as development of materials for schools, undergraduate programs, technical training, and public outreach; and research directed at identifying and quantifying the broad implications of nanotechnology for society, including social, economic, workforce, educational, ethical, and legal implications.

Under the 2007 NNI Strategic Plan, the Societal Dimensions PCA was divided into two separate PCAs: Environment, Health, and Safety, and Education and Societal Dimensions. Future PCA reporting will use the new eight PCA taxonomy. NSTC has retroactively reported FY2007 Societal Dimensions PCA spending in the new PCAs.[55] The NSET subcommittee characterizes the new PCAs as follows:[56]

Environment, Health, and Safety. This PCA addresses research primarily directed at understanding the environmental, health, and safety impacts of nanotechnology development and corresponding risk assessment, risk management, and methods for risk mitigation.

FY2008 funding for Environment, Health, and Safety rose to $58.6 million, up $10.3 million (21.3%) above the FY2007 level. A total of nine agencies funded work in this PCA in FY2008, including three agencies that did not have funding in this PCA in FY2008.[57] The President's FY2009 budget proposes $76.4 million in funding for this PCA, an increase of $17.8 million (30.4%) above the FY2008 level, led by increases at NIST (up $12 million) and EPA (up $4.7 million).[58]

Education and Societal Dimensions. This PCA addresses education- related activities such as development of materials for schools, undergraduate programs, technical training, and public communication, including outreach and engagement. Such activities include research directed at identifying and quantifying the broad implications of nanotechnology society, including social, economic, workforce, educational, ethical, and legal implications.

FY2008 funding for Education and Societal Dimensions fell slightly to $39 million, down $0.2 million (less than 1%) from the FY2007 level. The President's FY2009 budget proposes $40.7 million in funding for this PCA, an increase of $1 .7 million (4.4%) above the FY2008 level, accounted for entirely by an increase in NSF funding.

CENTERS, NETWORKS, AND USER FACILITIES

A key facet of the National Nanotechnology Initiative has been the development of an extensive infrastructure of interdisciplinary research and education centers, networks, and user facilities. The centers and user facilities are located at universities and federal laboratories across the country.

Centers and networks provide opportunities and support for multidisciplinary research among investigators from a variety of disciplines and research sectors, including academia, industry, and government laboratories. Such multidisciplinary research not only can lead to advances in knowledge, but also may foster relationships that further the development of basic research results into devices and other applications.

Many agencies support such centers. Examples of federal and federally supported centers include:

- The National Science Foundation has established university-based centers focused exclusively on nanotechnology, including 15 Nanoscale Science and Engineering Centers (NSEC), one Engineering Research Center, one Science and Technology Center, four Materials Research Science and Engineering Centers, two Nanoscale Science and Engineering Education Centers, and five Nanoscale Science and Engineering Networks.[59]
- The NIH has established more than 20 centers, including eight university-based Nanomedicine Development Centers; a Nanotechnology Characterization Laboratory, established by the National Cancer Institute (NCI), in partnership with NIST and the Food and Drug Administration; eight university-based Centers of Cancer Nanotechnology Excellence, established under the NCI's Alliance for Nanotechnology in Cancer initiative; and four university-based centers, established by the National Heart, Lung, and Blood Institute under its Program of Excellence in Nanotechnology.
- The Department of Defense supports two university-based nanotechnology research centers, as well as the Institute for Nanoscience at the Naval Research Laboratory.
- NASA has established three centers under its University Research, Engineering, and Technology Institute program.
- The Department of Energy has established five Nanoscale Science Research Centers (NSRCs) co-located with its national labs.
- NIST has established a Center for Nanoscale Science and Technology (CNST).
- NIOSH has established a Nanotechnology Research Center to conduct research into the application of nanoparticles and nanomaterials in occupational safety and health

and the implications of nanoparticles and nanomaterials for work-related injury and illness.

Many of the centers are designated as user facilities and are available to researchers not located at the center. User facilities are designed to allow outside researchers to take advantage of facilities, equipment, tools, and expertise. These shared resources provide researchers the opportunity to conduct research, characterize materials, and test products using equipment and facilities that their individual companies, universities, or organizations could not afford to acquire, support, or maintain. Conditions for user access vary by facility and agency. In general, users are not charged for pre-competitive, non-proprietary work leading to publication, and are charged on a cost-recovery basis for proprietary work. In some cases, the user facilities are located at federal government laboratories (e.g. the Department of Energy's five NSRCs, and the NIST CNST); other user facilities are located at universities and supported with federal funds (e.g. NSF's 13 university-based centers in the National Nanotechnology Infrastructure Network (NNIN)).

As mentioned earlier, the 21st Century Nanotechnology R&D Act of 2003 directed the establishment of two centers, the American Nanotechnology Preparedness Center and the Center for Nanomaterials Manufacturing. According to the NSET subcommittee, the requirement to establish the American Nanotechnology Preparedness Center was met by NSF's establishment of the Network for Nanotechnology in Society, comprised of centers at the University of California, Santa Barbara (with the participation of Harvard University and the University of South Carolina) and the University of Arizona.[60] These centers were funded under NSF's Nanoscale Science and Engineering Center (NSEC) program and did not include participation by any other NSET subcommittee agency.[61] The NSET subcommittee states that the requirement for establishing the Center for Nanomaterials Manufacturing was met by NSF's establishment of a National Nanomanufacturing Network (NNN) comprised of four NSECs. The Center for Integrated Hierarchical Manufacturing at the University of Massachusetts Amherst is the main node of the NNN.62 The NNN NSECs were established by NSF in collaboration with DOD and NIST, but exclusively with NSF funds.[63]

NNI Reports and Assessments

This section presents summaries of recent reports from the NSTC's Nanoscale Science, Engineering, and Technology Subcommittee and assessments conducted by the National Research Council and the President's Council of Advisors on Science and Technology.

NNI Reports

The NNI's coordinating body, the NSTC's Nanoscale Science, Engineering, and Technology Subcommittee, produces a variety of reports that serve to inform Congress and other key stakeholders on the initiatives' current activities, investments, and priorities.

The National Nanotechnology Strategic Plan (2007).[64] The 21st Century Nanotechnology R&D Act of 2003 (P.L. 108-153) requires the NSTC to develop an NNI strategic plan every three years. This plan is to guide the program's activities to meet the goals, priorities, and anticipated outcomes of the participating agencies. In addition, the act requires the triennial report to address how the program intends to move results out of the laboratory and into application for the benefit of society, its plan for long-term funding for interdisciplinary R&D, and the allocation of funding for interagency projects. The 2007 strategic plan is the first to follow external assessments by the National Academies and PCAST (operating as the NNAP) and seeks to incorporate the findings of these reviews. Of particular note:

- The 2007 National Nanotechnology Initiative Strategic Plan includes a new chapter on "High-Impact Application Opportunities and Critical Research Needs" possibly indicating an effort on the part of the Administration to move the NNI toward more directed research with commercial and societal benefits. Much of the early NNI work has been focused on basic research and mechanisms by which such research may produce economic and societal dividends. Seven years into the NNI, the program is under increasing scrutiny to deliver the promised benefits. While the plan does not establish R&D or application targets per se, this chapter illustrates tangible benefits that may be achieved by research supported under the NNI.
- In addition, the plan provides a matrix that identifies which agencies have a central or supporting role in each key application area. The plan also provides a series of nanotechnology application-specific vignettes on topics such as early detection of life-threatening disease, smarter computers, more energy-efficient transportation, and energy security.
- The 2007 National Nanotechnology Initiative Strategic Plan split the Societal Dimensions PCA into two separate PCAs: Environmental, Health, and Safety, and Education and Societal Dimensions. This change responds to increased Congressional and public attention to EHS needs. Some critics of the NNI had raised concerns that the inclusion of investments in education and other societal dimensions in the broader category obscured and artificially inflated the perception of investments in EHS R&D.
- The plan also identifies four areas of common interest across agencies that is to be the focus of future workshops: sensors and nanoelectronics, energy, fate and transport of nanomaterials, and medical and health applications.

The National Nanotechnology Initiative: Research and Development Leading to a Revolution in Technology and Industry, Supplement to the President's FY2008 Budget.[65] Each year the NSET subcommittee publishes a supplement to the President's annual budget request. The FY2008 NNI budget supplement provides a more detailed look at each agency's nanotechnology R&D budget request and a break-out of the prior, current, and requested year budgets for each PCA. In addition, the report provides data on agency funding for Small Business Innovation Research (SBIR) and Small Business Technology Transfer Research (STTR). The report describes proposed changes in agency R&D budgets, as well as in the balance of investments by PCA. Of particular note:

- The President's FY2008 budget proposed a $90.9 million (6.7%) increase in the overall NNI budget.
- Funding for EHS R&D in FY2007 rose to $47.8 million, a 26.8% increase over FY2006. The President's FY2008 request includes 58.6 million, a 22.6% increase over estimate FY2007 funding.
- The President's FY2008 budget request reflected a decline in spending on nanoscale devices and systems of $41.7 million (13.1%) below FY2007, led by a decline in DOD spending of $36.9 million (34.3%).

The National Nanotechnology Initiative: Environmental, Health, and Safety Research Needs for Engineered Nanoscale Materials[.66] This report documents the efforts of the NSET subcommittee's Nanotechnology Environmental and Health Implications (NEHI) working group to identify, prioritize, and implement research and other activities required for the responsible research and development of nanotechnology. The report is designed to help inform the research, risk assessment, and risk management activities of federal agencies and the private sector.

The report identifies priority research within five general research areas: instrumentation, metrology, and analytical methods; nanomaterials and human health; nanomaterials and the environment; health and environmental surveillance; and risk management methods.

The report identifies several next steps:

- prioritize research needs among those identified in the report;
- evaluate in greater detail the current NNI EHS research portfolio;
- perform a "gap analysis" of the NNI EHS research compared to the prioritized needs;
- coordinate and facilitate among the NNI agencies' research programs to address priorities; and
- establish a process for periodic review of progress and for updating research needs and priorities.

The report concludes that conducting EHS research in parallel with the development of nanomaterials and their applications will help to ensure the full, safe, and responsible realization of the promise of nanotechnology, and that coordination of research activities among NNI-participating agencies, as well as with industry and other governments, is necessary to expedite progress. In contrast, some NGOs have asserted the need for EHS research to precede the development of nanomaterials and nanotechnology applications.

Prioritization of Environmental, Health, and Safety Research Needs for Engineered Nanoscale Materials: An Interim Document for Public Comment.[67] This document is a follow-on to the Environmental, Health, and Safety Research Needs for Engineered Nanoscale Materials report (discussed above), incorporating public comments, refinements of the prioritization principles, and continued assessment of research needs. This report further identifies and defines five priorities within each of the five general categories of research needs established in the earlier document and presents the revised principles and the process used for this prioritization. The NEHI working group of the NSET subcommittee expects to use this report to evaluate the NNI's current EHS research portfolio, perform a gap analysis, and identify opportunities for interagency collaboration. The report stresses that the NSET subcommittee is "pursuing a dynamic, open, and transparent process in developing an NNI EHS research strategy" and invites continuing public input.

NNI Assessments

The 21st Century Nanotechnology R&D Act of 2003 (P.L. 108-153) requires periodic external reviews of the National Nanotechnology Program (NNP) by the National Research Council, an arm of the National Academies,[68] and the National Nanotechnology Advisory Panel.[69] In their first reviews, both institutions reviewed the NNI in its entirety, including the activities of those agencies that are not part of the NNP.

In general, these reviews concluded that the NNI has been successful so far and that its efforts are important to future U.S. technological leadership and commercial competitiveness. Both reports emphasize that much nanotechnology research is still in its very early stages and caution against expecting too much in the near term from this nascent technology. The reports also laud the cooperative efforts between the NNI and stakeholders in academia and industry and encourage increased interactions with industry, state and local economic developers, and, where appropriate, international partners.

A Matter of Size: Triennial Review of the National Nanotechnology Initiative.[70] This 2006 report presents the findings of the National Research Council's (NRC) first triennial review mandated by the 21st Century Nanotechnology R&D Act of 2003.

The NRC study concluded that the NNI has been successful in coordinating nanoscale efforts and interests across the federal government, in catalyzing cooperative R&D across a variety of scientific and engineering disciplines, and in opening a host of new scientific opportunities through its infrastructure and R&D investments. The NRC attributed much of this success to effective communication and coordination by the NSET subcommittee and the NNCO.

Recommendations. Here are the recommendations made by the NRC followed by a discussion of each.

NRC recommendation:

> the federal government [should] sustain investments in a manner that balances the pursuit of shorter-term goals with support for longer-term R&D and that ensures a robust supporting infrastructure, broadly defined. Supporting long-term research effectively will require making new funds available that do not come at the expense of much-needed ongoing investment in U.S. physical sciences and engineering research.

President Bush has expressed support for increasing federal R&D funding for the physical sciences and engineering, most notably in his American Competitiveness Initiative which includes nanotechnology investments. Yet, as Federal non- discretionary spending growth increases pressure on federal discretionary spending, finding new funds to support long-term nanotechnology research may need to come from other scientific disciplines.

NRC recommendation:

> the federal government [should] establish an independent advisory panel with specific operational expertise in nanoscale science and engineering; management of research centers, facilities, and partnerships; and interdisciplinary collaboration to facilitate cutting-edge research on and effective and responsible development of nanotechnology.

In July 2004, President George W. Bush implemented the provision of the 21st Century Nanotechnology Research and Development Act to "establish or designate a National Nanotechnology Advisory Panel" by issuing Executive Order 13349, which amends Executive Order 13226, designating the President's Council of Advisors on Science and Technology to serve as the NNAP.

The NRC's recommendation suggests that the President's designation of PCAST to serve as the legislatively mandated National Nanotechnology Advisory Panel is not fully adequate. Critics of the use of PCAST to serve as the NNAP maintain that the scope and depth of expertise needed to provide effective guidance on the NNI requires an independent panel of people with nanotechnology- and interdisciplinary-specific expertise and an undivided focus. Supporters of the use of PCAST for this function assert that a single advisory panel provides an integrated perspective, reduces unnecessary cost and management burdens, and that expertise can be added to the panel or accessed through non-member technical advisory groups.

NRC recommendation:

> federal agencies participating in the NNI, in consultation with the NNCO and the Office of Management and Budget, should continue to develop and enhance means for consistent tracking and reporting of funds requested, authorized, and expended annually. The current set of PCAs provides an appropriate initial template for such tracking.

It is difficult to assess and track funding for specific purposes within the NNI because the initiative is not centrally funded and operated. The NNI budget is an aggregation of the nanotechnology-related activities of the participating federal agencies. Congress funds the NNI-related R&D on an agency-by-agency basis, with responsibilities crossing many authorizing committees and appropriations subcommittees. Thus, while it is relatively straightforward to quantify an agency's nanotechnology budget, tracking all NNI investments related to a particular activity — EHS-related research, for example — is much more difficult. The PCAs serve to provide such a tracking mechanism. In addition, according to the 2007 National Nanotechnology Initiative Strategic Plan, the division of the Societal Dimensions PCA into two PCAs — Environmental, Health, and Safety; and Educational and Societal Dimensions — is intended to better understand and manage the NNI investment. Such a change indicates a level of flexibility that may enable the executive branch and Congress to more effectively manage and balance investments in discrete areas of the NNI.

NRC recommendation:

> the NSET Subcommittee [should] carry out or commission a study on the feasibility of developing metrics to quantify the return to the U.S. economy from the federal investment in

> nanotechnology R&D. The study should draw on the Department of Commerce's expertise in economic analysis and its existing ability to poll U.S. industry. Among the activities for which metrics should be developed and relevant data collected are technology transfer and commercial development of nanotechnology.

Few efforts have been made within the federal government to understand the economic impacts of the nation's investments in the NNI. Identification and tracking of data that could serve as an indicator of success in commercializing nanotechnology research or the effects on U.S. job creation or retention has not been formalized. To the extent that federal assessments of the economic contribution of and/or potential for nanotechnology products have occurred, they have not been performed with analytical rigor. Although the Commerce Department retains its economic analysis expertise, resident primarily in the Economics and Statistics Administration's Bureau of Economic Analysis, the Department's Technology Administration, which led Commerce's NNI activities and had government-wide responsibilities for technology transfer activities, was eliminated in August 2007.[71] Prior to its elimination, the Technology Administration contracted for two studies that could contribute to addressing this NRC recommendation: an analysis of barriers to nanotechnology commercialization performed by the University of Illinois at Springfield, and an analysis of innovation metrics conducted by the Alliance for Science and Technology Research in America (ASTRA). These reports are publicly available at Commerce Department websites.[72]

NRC recommendation:

> research on the environmental, health, and safety effects of nanotechnology [should] be expanded. Assessing the effects of engineered nanomaterials on public health and the environment requires that the research conducted be well-defined and reproducible, and that effective methods be developed and applied to (1) estimate the exposure of humans, wildlife, and other ecological receptors to source material; (2) assess effects on human health and ecosystems of both occupational and environmental exposure; and (3) characterize, assess, and manage the risks associated with exposure.

While the NRC asserts the need for additional EHS research, it does not quantify how much more is needed. Clayton Teague, director of the NNCO, has testified that the current level of investment in EHS research is adequate.[73] Many critics from academia, industry, and non-profit organizations have argued strongly that the NNI needs a greater level of investment in EHS research.[74] These critics argue from a variety of perspectives, including the need to:

- protect workers, human health, and the environment;
- create public faith and confidence in the safety of nanotechnology products;
- prevent a problem with one specific nanotechnology product from resulting in a loss of public support for all nanotechnology R&D; and
- create a predictable and stable regulatory environment.

This last factor is deemed by some as critical to fostering future nanotechnology investments.

NRC recommendation:

> the NSET Subcommittee [should] create a working group on education and the workforce that engages the Department of Education and Department of Labor as active participants.

The NSET subcommittee has sought, with limited success, greater involvement of the Departments of Education and Labor in the subcommittee's activities. An NSET subcommittee working group on education and the workforce has not yet been established.

With advocates promising the creation of many new jobs — some assert millions — as a result of global nanotechnology investments, some have expressed concern that the country must prepare students for nanotechnology research, engineering, and production jobs[75] Assessing which industries are likely to create such jobs, which skills will be needed, and in what timeframe are key challenges. If workers with nanotechnology-specific skills are needed and no workers are available domestically (U.S. citizens, resident aliens, or those in the United States on work visas), potential employers may opt to establish or move operations outside the United States to tap workers with those skills abroad. Conversely, if students are trained for jobs that do not emerge or do not emerge in the same timeframe as students are entering the job market, this investment is lost. In addition, potential students may be discouraged from pursuing future nanotechnology-related studies. Close coordination among the Departments of Commerce, Education, and Labor might help to align federal education and training efforts better with the labor market for nanotechnology workers.

The 21st Century Nanotechnology R&D Act also directed the NRC to address two other issues in its first triennial report: Is molecular self assembly feasible for manufacturing of materials and devices at the nanoscale? And, what are the needs for standards, guidelines, or strategies for ensuring the responsible development of nanotechnology?

Molecular Self-Assembly. Self-assembly is the process by which components (atoms, molecules, or more complex structures) form, without external control or direction, an organized structure. For example, water molecules dispersed in air in cold temperatures can self-assemble to form snowflakes. Our bodies act as self-assemblers, producing a variety of cells as needed (e.g. to repair a damage to the skin or produce new blood cells from added nutrients).

To what extent can molecular self-assembly be used as a tool for nanomanufacturing? On this issue, the NRC concluded that molecular self-assembly is feasible for the manufacture of simple materials and devices. However, for the manufacture of more sophisticated materials and devices, including complex objects produced in large quantities, the NRC found it unlikely that simple self-assembly processes will yield the desired results. One major barrier cited is the probability of error during assembly as a result of the systems' complexity.

Standards, Guidelines, and Strategies for Ensuring Responsible Development of Nanotechnology. The NRC concluded that it is not possible yet to make a rigorous assessment of the level of environmental and health risks posed by engineered nanomaterials and called for further development of risk assessment protocols. The NRC report also stated that the need for more EHS data requires an expanded research effort to complement dialog on these issues. In addition, until reproducible and well-characterized EHS data are available to inform the development of rigorous risk-based guidelines and best practices, the NRC found it prudent to recommend use of precautionary measures to protect the health and safety of workers, the public, and the environment. The NRC report also stressed that addressing the ethical and societal impacts of nanotechnology will require an integrated approach among scientists, engineers, social scientists, toxicologists, policymakers, and the public.

The National Nanotechnology Initiative at Five Years: Assessment and Recommendations of the National Nanotechnology Advisory Panel, President's Council of Advisors on Science and Technology.[76] This report presents the findings of the first biennial review of the NNI by the President's Council of Advisors on Science and Technology's, acting as the National Nanotechnology Advisory Panel, as mandated by the 21st Century Nanotechnology R&D Act of 2003. PCAST submitted its first report to the President on May 16, 2005, titled The National Nanotechnology Initiative at Five Years: Assessment and Recommendations of the National Nanotechnology Advisory Panel. The second report was due in 2007 but has yet to be completed.

The PCAST report finds that the United States is the acknowledged leader in nanotechnology R&D, but the U.S. leadership position is under increasing competitive pressure from growing public and private investments around the world. The report states that the federal investment in the NNI has been well-spent, the United States is well-positioned to maintain global leadership going forward, and continued robust funding is important for long-term U.S. economic well-being and national security. This assessment of the U.S. leadership position is founded not on sales, growth, or market share of commercial products — common measures of global competitiveness for established products — but rather on metrics that may serve as early indicators of potential innovation, such as the U.S. share of scientific publications and patents. The use of such metrics may not be universally accepted as predictive of leadership position. Technological leadership — or even leadership in innovation — does not ensure that the economic benefits from such leadership will accrue to the United States. Companies may choose to manufacture products or conduct other value-added activities outside the United States. If the assessment of national competitiveness is expanded to include the value-added activities and jobs generated or retained within the United States, then the metrics for assessing leadership might change.

The PCAST report acknowledges that there are potential environmental and health risks associated with nanotechnology, but finds that the NNI is directing appropriate attention and adequate resources to the research that will ensure the protection of the public and the environment. Nanotechnology products should not be immune from regulation, according to the report, but such regulation must be rational and based on science, not on perceived fears. The PCAST report states that strong communication exists among the NNI agencies responsible for research and regulation. The PCAST report contains four recommendations for the NNI:

PCAST recommendation:

> To further facilitate technology transfer from the lab to the marketplace, the NNI should expand its interaction with industry, increase federal-state coordination, and improve knowledge management of and access to NNI assets, such as user facilities and instrumentation.

The NSET subcommittee's National Innovation and Liaison with Industry (NILI) working group was established to facilitate NNI interactions with industry, and with state and local nanotechnology initiatives. The NILI working group's limited resources and agency participation have hindered its ability to conduct more extensive and sustained outreach.[77] Due to the structure and resource allocation of the NNI, the initiative's engagements with industry and with state and local initiatives are largely limited to single agency or laboratory interactions and to public engagement activities, such as speeches and information on the NNI website.[78]

PCAST recommendation:

> The NNI should continue its efforts to understand the possible toxicological effects of nanotechnology and where harmful human or environmental effects are proven, pertinent federal agencies should apply appropriate regulatory mechanisms. There should be strong interagency and international collaboration on this issue to eliminate unnecessary duplication of research efforts and to ensure wide dissemination of information. Since exposure to nanomaterials is most likely to occur during the manufacturing process, research on potential hazards associated with workplace exposure must be given the highest priority.

With respect to collaboration on EHS issues, the NSET subcommittee's National Environmental and Health Implications (NEHI) working group is the primary EHS coordination mechanism for participating NNI agencies. The Global Issues in Nanotechnology (GIN) working group works with the NEHI working group on international collaboration on EHS issues. The NIOSH has published several documents addressing concerns about workplace exposure to nanoparticles. Approaches to Safe Nanotechnology: An Information Exchange With NIOSH was intended to provide the best currently available knowledge on nanoparticle toxicity and control and to solicit input from the stakeholder community.[79] Progress Toward Safe Nanotechnology in the Workplace details the work of NIOSH's Nanotechnology Research Center from 2004 through 2006.[80] In December 2007, NIOSH released interim guidance concerning the medical screening of workers potentially exposed to engineered nanoparticles during the manufacture and industrial use of nanomaterials. The NIOSH says that the document is intended to "generate discussion, fill the current knowledge gap, and provide interim recommendations until further scientific information becomes available."[81] The NIOSH is currently seeking public comment on this guidance.

With respect to regulatory issues associated with nanotechnology, see CRS Report RL34332, Engineered Nanoscale Materials and Derivative Products: Regulatory Challenges, by Linda-Jo Schierow.

PCAST recommendation:

> The NNI should establish relationships with the Department of Education and Department of Labor to develop education and training systems to support the Nation's technical proficiency in areas related to nanotechnology.

The PCAST report's recommendation is similar to the recommendation made by the NRC and is discussed earlier in this paper.

PCAST recommendation:

> The NNI must support research aimed at understanding the societal implications of nanotechnology — including ethical, economic and legal implications — and must actively work to inform the public about nanotechnology. The NNI should continue to confront societal issues in an open, straightforward, and science-based manner.

Some critics of the NNI hold deep reservations about the ethical, economic, and legal implications of nanotechnology. Some of these concerns are common to many technologies, such as the allocation of risk and benefit during manufacturing. For example, a neighborhood located near a production facility may bear risks associated with exposure to the byproducts (or products) of manufacturing, while gaining few of the benefits. Concerns about possible adverse effects of nanoscale particles on human health and the environment resulting from their small particle size and unique characteristics may result in increased attention to such costs and benefits with respect to nanoscale material production. Currently, nanotechnology EHS risks are unknown and may be acute or pose no more risk than other manufacturing processes.

Privacy rights are another issue associated with the products of nanotechnology. Nanotechnology may enable the production of highly sensitive, inexpensive sensors that could be deployed ubiquitously in commercial and public settings. While these sensors may allow check-out-free purchases from stores, or monitor the environment for toxic substances, critics argue that they could also impinge on the privacy rights of individuals if, for example, the sensors could detect chemicals related to the use of tobacco, alcohol, or illegal substances without the permission of the individual. Such information might be later applied in law enforcement, life insurance, health insurance, or employment decisions.[82] Others express concern that the economically disadvantaged and less educated — both individuals and nations — might be unable to take part in the benefits that nanotechnology products could offer.[83]

On the legal front, innovations in nanotechnology are already presenting unique challenges to the U.S. Patent and Trademark (USPTO). For example, U.S. case law generally prohibits patenting where the sole element of novelty is a change in size, the characteristic most obviously associated with nanotechnology.[84] In addition, many nanotechnology innovations involve multiple disciplines. Since the USPTO structure for examining patents is discipline-based, an examiner may not have all of the requisite expertise for the examination, affecting both their ability to conduct the examination, and the speed at which it can be done. USPTO also has acknowledged the need to accelerate the speed of nanotechnology-related patent applications. According to John Doll, Commissioner of Patents, the agency is hampered in its ability to recruit and retain patent examiners with the requisite skills to handle nanotechnology patents given the "more generous offers [patent examiners get] from the private sector."[85] Doll said that efforts have been made to improve hiring and retention at USPTO, and that a new processes has been established allowing an accelerated examination of applications.[86]

Nanotechnology Legislation in the 110th Congress S. 1199/H.R. 2436 — Nanotechnology in the Schools Act

S. 1199, the Nanotechnology in the Schools Act, was introduced on April 24, 2007. Companion legislation, H.R. 2436, was introduced on May 22, 2007. The bill seeks to strengthen the capacity of U.S. secondary schools and institutions of higher education to prepare students for careers in nanotechnology by providing grants to those schools and institutions.

Under the legislation, the Director of the National Science Foundation is directed to establish a nanotechnology in the schools program. This program would award grants of not more than $150,000 to public or charter secondary schools offering advanced science courses and to institutions of higher education. These grants would be for the purchase of nanotechnology equipment and software and the provision of nanotechnology education to students and teachers. Institutions must provide non-federal matching funds of 25% of the grant amount, but this provision can be waived for institutions with endowments of $5 million or less.

The bill authorizes $15 million for the program in FY2008, and such sums as may be necessary for FY2009 through FY201 1. S. 1199 was referred to the Senate Committee on Health, Education, Labor, and Pensions. H.R. 2436 was referred to the House Science and Technology Committee's Subcommittee on Research and Science Education.

S. 1547 — National Defense Authorization Act for Fiscal Year 2008 (Incorporating the Provisions of S. 1425)

S. 1425 was introduced on May 17, 2007. The purpose of S. 1425 was to enhance the Department of Defense's nanotechnology R&D program. The provisions of S. 1425 were later incorporated in S. 1547, the National Defense Authorization Act for Fiscal Year 2008, as section 255. The bill would amend language in the Bob Stump National Defense Authorization Act for Fiscal Year 2003 relating to the defense nanotechnology R&D program. Section 255 of S. 1547 would:

- revise the nanotechnology R&D program purposes;
- replace the Director of Defense Research and Engineering with the Under Secretary of Defense for Acquisition, Technology, and Logistics as the program administrator responsible for overseeing coordination of the nanotechnology R&D program with other departments and agencies participating in the NNI;
- outline program activities, including the establishment of research priorities; development of a strategic plan for defense nanotechnology R&D that is integrated with the strategic plan for the NNI; and establishment of a strategy for transitioning research results into products needed by DOD, including support for development of nanomanufacturing capabilities and a nanotechnology defense industrial base; and
- extend program reporting requirements through 2013.

The bill also requires a report from the Comptroller General to the congressional defense and appropriations committees on progress made by DOD in achieving the purposes of the program. S. 1425 was referred to the Senate Committee on Armed Services. The provisions of this bill have been incorporated in section 255 of S. 1547, the Department of Defense Authorization Act for Fiscal Year 2008.

H.R. 3235 — Nanotechnology Advancement and New Opportunities Act

H.R. 3235, the Nanotechnology Advancement and New Opportunities Act, was introduced on July 31, 2007. The purpose of the bill is to ensure the development and responsible stewardship of nanotechnology. The bill would:

- establish a $100 million Nanomanufacturing Investment Partnership at the Department of Commerce to work with private investors to advance the commercialization of nanomanufacturing technologies and to increase the commercial application of federally supported research results;
- establish a tax credit of up to $10 million for the purchase of stock in qualified nanotechnology companies;
- establish a grant program within the DOC to support the establishment and development of nanotechnology incubators;
- authorize $10 million for the NSF to establish a center for the development of computer-aided design tools for nanotechnology applications;
- authorize an annual appropriation of $30 million for the DOE to conduct a grant program for nanotechnology research to address the need for clean, cheap, renewable energy;
- authorize an annual appropriation of $30 million for the EPA for a grant program for nanotechnology research to address technologies for the remediation of pollution and other environmental protection technologies;
- authorize an annual appropriation of $30 million for the DHS to conduct a grant program for nanotechnology research to address the need for sensors and materials related to homeland security needs;
- authorize an annual appropriation of $30 million for the DHHS to conduct a grant program for nanotechnology research to address health-related applications;
- require the Director of the NNCO to prepare a report to Congress on a nanotechnology research strategy for government and industry that will ensure the development and responsible stewardship of nanotechnology;
- provide a tax credit of 50% for nanotechnology education and training expenses for businesses and individuals;
- authorize an annual appropriation of $15 million for FY2008 through FY20 11 for the NSF to conduct a grant program for the development of curriculum materials for interdisciplinary nanotechnology courses at institutions of higher education;
- direct the NSF to establish, through its Advanced Technological Education program, a program to encourage manufacturing companies to enter into partnerships with

occupational training centers for the development of training to support nanomanufacturing; and

- direct the Secretary of Energy to submit a report to Congress containing a strategy for increasing interaction among scientists and engineers at DOE national laboratories and the informal science education community to prepare appropriate exhibits for school age children and the general public.

On July 31, 2007, H.R. 3235 was referred to the House Science and Technology Committee's Subcommittee on Research and Science Education; the House Ways and Means Committee; the House Energy and Commerce Committee's Subcommittee on Commerce, Trade and Consumer Protection; and the House Homeland Security Committee's Subcommittee on Emerging Threats, Cybersecurity, and Science and Technology.

S. 1372 — Nanotechnology Infrastructure Enhancement Act

S. 1372, the Nanotechnology Infrastructure Enhancement Act, was introduced on May 11, 2007. The purpose of the bill is to improve the national nanotechnology infrastructure by establishing a Nanoscale Science and Engineering Center or node on the National Nanotechnology Infrastructure Network located in an EPSCoR (Experimental Program to Stimulate Competitive Research) state.[87] The bill would:

- require the Director of the NSF to establish a geographically diverse, interdisciplinary Center for Nanotechnology Research and Engineering focused on either the science and engineering of manufacturing at the nanoscale in multiple dimensions or nanotechnology for sustainable energy, water, agriculture, and the environment;
- allow the center to be a Nanoscale Science and Engineering Center or a National Nanotechnology Infrastructure Network node;
- require that the center shall include a lead academic institution located in an EPSCoR state and at least one additional academic institution located in a second EPSCoR state, while permitting the center to include other institutions within or outside of the United States; and
- require the center to conduct state-of-the-art research on nanomanufacturing; collaborate with other NSF grantees, and with grantees from other federal agencies, working on nanomanufacturing; share resources with the programs of other NSF grantees for the purpose of mutual advantage; and work toward a nanomanufacturing network that encourages extensive industrial collaboration.

Concluding Observations

Many expect nanotechnology to bring significant economic and societal returns. The United States was the first government to launch a national-level nanotechnology program and has invested more than any other nation. As a result of this focus and these sustained investments, many experts believe that the United States enjoys a technological leadership position in nanotechnology. Other nations are investing heavily and some industrialized and emerging economies have formidable capabilities in nanotechnology. Assessments of the National Nanotechnology Initiative have concluded that the effort is well-managed and has been successful in achieving its objectives so far. However, these assessments have recognized that the NNI faces a variety of challenges in ensuring that the full promise of nanotechnology is realized and that the United States remains the global leader in nanoscale science, engineering, and technology.

Congress may choose to address some or many of the issues addressed in the body of this report in the course of deliberation on the reauthorization of the 21st Century Nanotechnology R&D Act of 2003 or, alternatively, in separate legislation.

The 21st Century Nanotechnology R&D Act's funding authorizations extend through FY2008. Action is being considered in both the House and Senate on reauthorization of the program. Possible topics for consideration in the reauthorization process include budget authorization levels for the covered agencies; R&D funding levels, priorities, and balance across the program component areas; administration and management of the NNI; translation of research results and early- stage technology into commercially viable applications; environmental, health, and safety issues; ethical, legal, and societal implications; education and training for the nanotechnology workforce; metrology, standards, and nomenclature; public understanding; and international dimensions. Consideration may also be given to the establishment of an independent review panel and to coordination of the timing for the NNAP assessment, the NRC assessment, and the NSET subcommittee's strategic plan for the NNI.

Appendix A. Selected Reports on the National Nanotechnology Initiative

Reports of the Nanoscale Science, Engineering, and Technology Subcommittee of the National Science and Technology Council

The National Nanotechnology Initiative Strategic Plan. December 2007.

The National Nanotechnology Initiative: Research and Development Leading to a Revolution in Technology and Industry, Supplement to the President's FY2008 Budget. July 2007.

The National Nanotechnology Initiative: Environmental, Health, and Safety Research Needs for Engineered Nanoscale Materials. September 2006.

The National Nanotechnology Initiative: Research and Development Leading to a Revolution in Technology and Industry, Supplement to the President's FY2007 Budget. July 2006.

The National Nanotechnology Initiative: Research and Development Leading to a Revolution in Technology and Industry, Supplement to the President's FY2006 Budget. March 2005.

The National Nanotechnology Initiative Strategic Plan. December 2004. Nanotechnology in Space Exploration. August 2004.

Nanoscience Research for Energy Needs. Report from a workshop held in March 2004.

Instrumentation and Metrology for Nanotechnology. Report from a workshop held in January 2004.

Nanotechnology: Societal Implications-Maximizing Benefits for Humanity. Report from a workshop held in December 2003.

Nanobiotechnology. Report from a workshop held in October 2003.

Regional, State, and Local Initiatives in Nanotechnology. Report from a workshop held in September-October 2003.

National Nanotechnology Initiative: Research and Development Supporting the Next Industrial Revolution, Supplement to the President's FY2004 Budget. August 2003.

National Nanotechnology Initiative: The Initiative and Its Implementation Plan, Detailed Technical Report Associated with the Supplemental Report to the President's FY2003 Budget. June 2002.

National Nanotechnology Initiative: The Initiative and Its Implementation Plan, Detailed Technical Report Associated with the Supplemental Report to the President's FY2001 Budget. July 2000.

Report of the Interagency Working Group on Nanoscience, Technology, and Engineering (NSET Subcommittee Predecessor)

Nanotechnology: Shaping the World Atom by Atom. 1999.

Agency Reports

NIOSH Update: NIOSH Draft Offers Interim Guidance on Medical Screening of Workers Potentially Exposed to Engineered Nanoparticles, National Institute for Occupational Safety and Health. December 2007.

Progress Toward Safe Nanotechnology in the Workplace, National Institute for Occupational Safety and Health. June 2007.

Approaches to Safe Nanotechnology in the Workplace, National Institute for Occupational Safety and Health. July 2006.

Nanoscale Science, Engineering, and Technology in DOE's Office of Basic Energy Sciences, U.S. Department of Energy. February 2003.

External Reviews

A Matter of Size: Triennial Review of the National Nanotechnology Initiative, National Research Council. 2006.

The National Nanotechnology Initiative at Five Years: Assessment and Recommendations of the National Nanotechnology Advisory Panel, President's Council of Advisors on Science and Technology (acting as the National Nanotechnology Advisory Panel). May 2005.

Small Wonders, Endless Frontiers: A Review of the National Nanotechnology Initiative, National Research Council. June 2002.

Appendix B. List of NNI and Nanotechnology-Related Acronyms

ASTRA	Alliance for Science and Technology Research in America
CNST	Center for Nanoscale Science and Technology
CS	Committee on Science
CT	Committee on Technology
CSREES	Cooperative State Research, Education, and Extension Service
DHHS	Department of Health and Human Services
DHS	Department of Homeland Security
DOC	Department of Commerce
DOD	Department of Defense
DOE	Department of Energy
DOJ	Department of Justice
DOT	Department of Transportation
EHS	Environmental, health, and safety
ELSI	Ethical, legal, and societal implications
EPA	Environmental Protection Agency
EOP	Executive Office of the President
EPSCoR	Experimental Program to Stimulate Competitive Research
FHWA	Federal Highway Administration
GIN	Global Issues in Nanotechnology working group
ISO	International Standards Organization
IWGN	Interagency Working Group on Nanotechnology
NASA	National Aeronautics and Space Administration
NCI	National Cancer Institute
NEHI	National Environmental and Health Implications working group
NGO	Non-governmental organization
NIH	National Institutes of Health
NILI	National Innovation and Liaison with Industry working group
NIOSH	National Institute of Occupational Safety and Health
NIST	National Institute of Standards and Technology
NNAP	National Nanotechnology Advisory Panel
NNCO	National Nanotechnology Coordination Office
NNI	National Nanotechnology Initiative
NNIN	National Nanotechnology Infrastructure Network
NNN	National Nanomanufacturing Network
NNP	National Nanotechnology Program
NPEC	Nanotechnology Public Engagement and Communications working group
NRC	National Research Council
NSET	Nanoscale Science, Engineering, and Technology subcommittee

NSF	National Science Foundation
NSEC	Nanoscale Science and Engineering Center
NSRC	Nanoscale Science Research Centers
NSTC	National Science and Technology Council
OECD	Organization for Economic Cooperation and Development
OMB	Office of Management and Budget
OSTP	Office of Science and Technology Policy
PCA	Program Component Areas
PCAST	President's Council of Advisors on Science and Technology
R&D	Research and development
SBIR	Small Business Innovation Research
STTR	Small Business Technology Transfer Research
TSA	Transportation Security Administration
USDA	U.S. Department of Agriculture
USPTO	U.S. Patent and Trademark Office

REFERENCES

[1] In the context of the NNI and nanotechnology, the nanoscale refers to a dimension of 1 to 100 nanometers (see box below).

[2] While extensive R&D has been, and continues to be, conducted to understand and harness the properties of individual atoms, this is not the domain of nanotechnology.

[3] Profiting From International Nanotechnology, Lux Research, December 2006.

[4] The National Nanotechnology Initiative at Five Years: Assessment and Recommendations of the National Nanotechnology Advisory Panel, President's Council of Advisors on Science and Technology, May 2005.

[5] "Nanotechnology Moves from Discovery to Commercialization," press release, Lux Research, November 20, 2007. [http://www.luxresearchinc.com/press/ 2007-lux-research-nanotech-report-5 .pdf]

[6] Sizing Nanotechnology 's Value Chain, Lux Research, 2004.

[7] Halfway to the Trillion Dollar Market: A Critical Review of the Diffusion of Nanotechnologies, Cientifica, 2007. [http://www.cientifica.eu/files/Whitepapers/ A%20Reassessment%20of%20the%20Trillion%20WP.pdf]

[8] While views vary on how to calculate nanotechnology's contribution to these products, the consensus is that nanotechnology is likely to have a significant economic impact and transformative effect on many industries.

[9] National Cancer Institute website. [http://nano.cancer.gov/resource_center/tech_ backgrounder.asp]

[10] Ibid.

[11] Aslan, Kadir; Lakowicz, Joseph R.; and Geddes, Chris D. "Nanogold plasmon resonance- based glucose sensing. Wavelength-ratiometric resonance light scattering," Analytical Chemistry, 2005, Vol. 77. Strategic Plan for Pediatric Urology, National Institute of Diabetes and Digestive and Kidney Disease, National Institutes of Health, Department of Health and Human Services, February 2006.

[12] Risbud, Aditi. "Fruit of the Nano Loom," Technology Review, February 2006.

[13] Nanoscience Research for Energy Needs, Nanoscale Science, Engineering, and Technology Subcommittee, National Science and Technology Council, The White House, December 2004.

[14] Risbud, Aditi. "Cheap Drinking Water from the Ocean," Technology Review, June 2006.

[15] Selko, Adrienne. "New Nanotechnology-Based Coatings Are Energy Efficient and Environmentally Sound," Industry Week, August 22, 2007. "Tomorrow's Green Nanofactories," Science Daily, July 11, 2007.

[16] 16 National Nanotechnology Initiative — Leading to the Next Industrial Revolution, Interagency Working Group on Nanoscience, Engineering, and Technology, National Science and Technology Council, The White House. [http://www.ostp.gov/NSTC/html/iwgn/iwgn.fy0 1budsuppl/nni.pdf]

[17] 21st Century Agriculture: A Critical Role for Science and Technology, U.S. Department of Agriculture, June 2003; and Nanoscale Science and Engineering for Agriculture and Food Systems: Draft Report of the National Planning Workshop to the Cooperative State Research, Education, and Extension Service of the U.S. Department of Agriculture, July 2003.

[18] Nanotechnology in Space Exploration, Nanoscale Science, Engineering, and Technology Subcommittee, National Science and Technology Council, The White House, August 2004.

[19] Nanotechnology and the Environment, Nanoscale Science, Engineering, and Technology Subcommittee, National Science and Technology Council, The White House, May 2003.

[20] Proceedings of the U.S. Environmental Protection Agency Workshop on Nanotechnology for Site Remediation, U.S. Environmental Protection Agency, October 2005.

[21] "National Nanotechnology Initiative: Leading to the Next Industrial Revolution," press release, The White House, January 21, 2000. [http://clinton4.nara.gov/WH/New/html/2000012 1_4.html]; and "Steering the technology that will redefine life as we know it," Industrial Biotechnology, Vol. 1, No. 3, Fall 2005. [http://www.nsf.gov/crssprgm/nano/reports/mcr_ind_ biotech_interview. pdf]

[22] Prior to P.L. 108-153, the Bob Stump Defense Authorization Act for Fiscal Year 2003 (P.L. 107-314) required DOD to "provide for interagency cooperation and collaboration on nanoscale research and development." The NSET subcommittee is a subcommittee of the NSTC Committee on Technology.

[23] Metrology is the science of measurement, including the equipment and processes used to produce a measurement.

[24] The Department of Commerce characterizes emerging and enabling technologies as those that "offer a wide breadth of potential application and form an important technical basis for future commercial applications." (ATP Rule, 15 C.F.R. Part 295).

[25] The term "valley of death" is used by business executives, economists, and venture capitalists to describe the development gap that often exists between a laboratory discovery and the market's willingness to invest to advance the discovery to a final commercial product. This gap occurs due to a variety of issues, such as technical risk, market uncertainty, and likelihood of obtaining an adequate return on investment.

[26] Crews, Clyde Wayne, Jr., "Washington's Big Little Pork Barrel: Nanotechnology," Cato Institute website, May 29, 2003. [http://www.cato.org/pub_display.php?pub_id=3 110]

[27] Testimony of Andrew Maynard, Chief Science Advisor, Project on Emerging Nanotechnologies, Woodrow Wilson International Center for Scholars, "Research on Environmental and Safety Impacts of Nanotechnology: Current Status of Planning and Implementation under the National Nanotechnology Initiative," hearing, Subcommittee on Research and Science Education, House Committee on Science and Technology, October 31, 2007.

[28] "NGOs urge precautionary principle in use of nanomaterials," EurActiv.com, June 14, 2007. [http : //www.euractiv.com/en/environment/ngos-urge-precautionary-prin ciple-use-nanomaterials/article-1 64619] Sass, Jennifer. "Nanotechnology and the Precautionary Principle," presentation, Natural Resources Defense Council, 2006. [http://docs.nrdc.org/health/hea_06121402a.pdf] The precautionary principle has been used in other countries on some issues. For example, the Biosafety Protocol to the 1992 Convention on Biological Diversity incorporates provisions applying the precautionary principle to the safe handling, transfer, and trade of genetically modified organisms. For further information, see CRS Report RL30594, Biosafety Protocol for Genetically Modified Organisms: Overview, by Alejandro E. Segarra and Susan R. Fletcher.

[29] "No Small Matter II: The Case for a Global Moratorium — Size Matters!," Occasional Paper Series, ETC Group, April 2003. [http://www.etcgroup.org/upload/publication/ pdf_file/1 65]

[30] The National Nanotechnology Initiative Strategic Plan, Nanoscale Science, Engineering, and Technology Subcommittee, National Science and Technology Council, The White House, December 2004.

[31] Ibid.

[32] National Nanotechnology Initiative website. [http://www.nano.gov/html/about/ history.html]

[33] "National Nanotechnology Initiative: Leading to the Next Industrial Revolution," press release, The White House, January 21, 2000. [http://clinton4.nara.gov/WH/New/html/ 2000012 1_4.html]; and National Nanotechnology Initiative website. [http://www.nano.gov/ html/about/history.html]

[34] In its January 21, 2001 press release, "National Nanotechnology Initiative: Leading to the Next Industrial Revolution," announcing the establishment of the NNI, the White House identified only six participating agencies — NSF, DOD, DOE, NIST, NASA, and NIH. Subsequently, EPA and DOJ reported nanotechnology R&D funding in FY2001, bringing the total number of agencies funding nanotechnology R&D in FY200 1 to eight.

[35] NNI participants include agencies that either conduct or provide funding for nanotechnology R&D, as well as agencies with missions that may affect the development, commercialization, and use of nanotechnology. For example, in the latter case, the Food and Drug Administration may regulate (or not regulate) nanotechnology products, the U.S. Patent and Trademark Office's (USPTO) treatment of nanotechnology-related patents may affect the value of the underlying intellectual property, and the execution of the missions of the Departments of Education and Labor could affect the preparedness of the U.S. workforce for emerging nanotechnology jobs.

Some nanotechnology R&D agencies may also have non-R&D missions related to nanotechnology. For example, EPA conducts and funds R&D but also has a regulatory mission that could affect nanotechnology research, development, production, use, and/or disposal.

[36] U.S. Congress. 2003. 21st Century Nanotechnology Research and Development Act. P.L. 108-153. 15 U.S.C. 7501. 108 Cong., December 3.

[37] Ibid.

[38] FY2003 funding attributed to DHS for the purpose of this calculation is based on nanotechnology R&D appropriations received by the Department of Transportation's Transportation Security Administration (TSA). TSA was transferred to DHS in the Homeland Security Act of 2002 (P.L. 107-296) which was enacted after the start of FY2003.

[39] The National Nanotechnology Initiative Strategic Plan, Nanoscale Science, Engineering, and Technology Subcommittee, National Science and Technology Council, The White House, December 2004.

[40] The agencies that participate in the NSET subcommittee comprise the NNI. NSETsubcommittee members include Bureau of Industry and Security, Department of Commerce;Consumer Product Safety Commission; Cooperative State Research, Education, andExtension Service, Department of Agriculture; Department of Defense; Department ofEducation; Department of Energy; Department of Homeland Security; Department ofJustice; Department of Labor; Department of State; Department of Transportation;Department of the Treasury; Environmental Protection Agency; Food and DrugAdministration; Forest Service, Department of Agriculture; Intelligence TechnologyInnovation Center; International Trade Commission; National Aeronautics and SpaceAdministration; National Institutes of Health, U.S. Department of Health and HumanServices; National Institute for Occupational Safety and Health, Center for Disease Control, Department of Health and Human Services; National Institute of Standards and Technology, Department of Commerce; National Science Foundation; Nuclear Regulatory Commission; U.S. Geological Survey; and the U.S. Patent and Trademark Office, Department of Commerce. The Department of Commerce's Technology Administration was a participating agency in the NNI until its elimination in August 2007 under the America COMPETES Act (P.L. 110-69).

[41] The National Nanotechnology Initiative: Research and Development Leading to a Revolution in Technology and Industry-Supplement to the President's FY2008 Budget, Nanoscale Science, Engineering, and Technology Subcommittee, National Science and Technology Council, The White House, July 2007.

[42] Testimony of E. Clayton Teague, Director, National Nanotechnology Coordinating Office, Research on Environmental and Safety Impacts of Nanotechnology: Current Status of Planning and Implementation under the National Nanotechnology Initiative, hearing, Subcommittee on Research and Science Education, House Committee on Science and Technology, October 31, 2007.

[43] The Industrial Research Institute is an association of companies and federally funded laboratories with the mission of improving R&D capabilities through the development and dissemination of best practices.

[44] National Nanotechnology Initiative website, [http://www.nano.gov].

[45] The eight agencies that are signatories to the memorandum of understanding are NSF, DOD, DOE, NIH, NIST, NASA, EPA, and DOT.

[46] The National Nanotechnology Initiative Strategic Plan, Nanoscale Science, Engineering, and Technology Subcommittee, National Science and Technology Council, The White House, December 2004.

[47] The National Nanotechnology Initiative Strategic Plan, Nanoscale Science, Engineering, and Technology Subcommittee, National Science and Technology Council, The White House, December 2007.

[48] The National Nanotechnology Initiative: FY2009 Budget & Highlights, Nanoscale Science, Engineering, and Technology Subcommittee, National Science and Technology Council, The White House, February 2008.

[49] Ibid.

[50] Ibid.

[51] Ibid.

[52] Top-down processes are those that achieve design features by removing material from a larger block of material; bottom-up processes begin with smaller building blocks (atoms or molecules) and achieve design features by putting them together, possibly using self-assembly.

[53] The National Nanotechnology Initiative: FY2009 Budget & Highlights, Nanoscale Science, Engineering, and Technology Subcommittee, National Science and Technology Council, The White House, February 2008.

[54] Ibid.

[55] The National Nanotechnology Initiative: FY2009 Budget & Highlights, Nanoscale Science, Engineering, and Technology Subcommittee, National Science and Technology Council, The White House, February 2008.

[56] The National Nanotechnology Initiative Strategic Plan, Nanoscale Science, Engineering, and Technology Subcommittee, National Science and Technology Council, The White House, December 2004.

[57] The three agencies reporting FY2008 funding in the Environment, Health, and Safety PCA that reported no funding for this PCA in FY2007 are DOD, DOE, and NASA.

[58] The National Nanotechnology Initiative: FY2009 Budget & Highlights, Nanoscale Science, Engineering, and Technology Subcommittee, National Science and Technology Council, The White House, February 2008.

[59] In addition, 18 other Materials Research Science and Engineering nanotechnology-related research as part of their overall efforts.

[60] Private telephone communication between CRS and NSTC staff, January 31, 2008.

[61] Private email communication between CRS and NSF staff, January 31, 2008.

[62] Private telephone communication between CRS and NSTC staff, January 31, 2008.

[63] Private e-mail communication between CRS and NSF staff, January 31, 2008.

[64] The National Nanotechnology Strategic Plan, Nanoscale, Science, Engineering, and Technology Subcommittee, National Science and Technology Council, The White House, December 2007.

[65] The National Nanotechnology Initiative: Research and Development Leading to a Revolution in Technology and Industry, Supplement to the President's FY2008 Budget, Nanoscale, Science, Engineering, and Technology Subcommittee, National Science and Technology Council, The White House, July 2007.

[66] The National Nanotechnology Initiative: Environmental, Health, and Safety Research Needs for Engineered Nanoscale Materials, Nanoscale, Science, Engineering, and Technology Subcommittee, National Science and Technology Council, The White House, September 2006.

[67] Prioritization of Environmental, Health, and Safety Research Needs for Engineered Nanoscale Materials: An Interim Document for Public Comment, Nanoscale, Science, Engineering, and Technology Subcommittee, National Science and Technology Council, The White House, August 2007.

[68] The National Research Council, the National Academy of Sciences, the National Academy of Engineering, and the Institute of Medicine are part of a private, non-profit institution established under a congressional charter. They are collectively referred to as the National Academies.

[69] P.L. 108-153 directs the President to "establish or designate a National Nanotechnology Advisory Panel." In July 2004, President Bush designated the President's Council of Advisors on Science and Technology to serve as the NNAP by issuing Executive Order 13349, Amending Executive Order 13226 To Designate the President's Council of Advisors on Science and Technology To Serve as the National Nanotechnology Advisory Panel.

[70] A Matter of Size: Triennial Review of the National Nanotechnology Initiative, National Research Council, 2006.

[71] The Technology Administration was eliminated in the America COMPETES Act (P.L. 110-69).

[72] Barriers to Nanotechnology Commercialization, College of Business Management, University of Illinois at Springfield, September 2007. [http://www.osec.doc.gov/Report-Barriers%20to%20Nanotechnology%20Commercialization.pdf] Innovation Vital Signs Project, Alliance for Science and Technology Research in America, July 2007. [http://www.ntis.gov/ta_reports/Report-InnovationVitalSigns.pdf]

[73] Testimony of Clayton Teague, director of the NNCO, hearing, "Environmental and Safety Impacts of Nanotechnology: What Research is Needed?" House Committee on Science, November 17, 2005. [http://commdocs.house.gov/committees /science/hsy 24464.000/ hsy24464_0.HTM]

[74] Ibid.

[75] Phillip J. Bond, Under Secretary for Technology, U.S. Department of Commerce, remarks, "Nanotechnology: Economic Opportunities, Societal and Ethical Challenges," NanoCommerce 2003, December 9, 2003. [http://www.technology.gov/Speeches/PJB_03 1209.htm] Sizing Nanotechnology 's Value Chain, Lux Research, October 2004.

[76] The National Nanotechnology Initiative at Five Years: Assessment and Recommendations of the National Nanotechnology Advisory Panel, President's Council of Advisors on Science and Technology, May 2005.

[77] Private telephone and e-mail communication with Sean Murdock, executive director of the NanoBusiness Alliance, February 4, 2008.

[78] Ibid.

[79] Approaches to Safe Nanotechnology: An Information Exchange with NIOSH, National Institute for Occupational Safety and Health, July 2006.

[80] Progress Toward Safe Nanotechnology in the Workplace, National Institute for Occupational Safety and Health, June 2007.

[81] NIOSH Update: NIOSH Draft Offers Interim Guidance on Medical Screening of Workers Potentially Exposed to Engineered Nanoparticles, National Institute for Occupational Safety and Health, December 13, 2007.

[82] Moore, Fiona M., "Implications of Nanotechnology Applications: Using Genetics as a Lesson," Health Law Review, Vol. 10, No. 3, 2002. [http://www.law.ualberta.ca/centres /hli/ pdfs/hlr/v1 0_3/1 0.3moorefrm.pdf]

[83] Smith, Richard H.,"Social, Ethical, and Legal Implications of Nanotechnology," Societal Implications of Nanoscience and Nanotechnology (The Netherlands:Kluwer Academic Publishers, 2001).

[84] Nanotechnology Patents: Issues for Nanotechnology Inventions, Dorsey and Whitney, LLP, May 9, 2005.

[85] Doll, John, Commissioner of Patents, U.S. Patent and Trademark Office, Letter to the Editor, Small Times, April 23, 2007. [http://www.smalltimes.com/Articles/Article_ Display.cfm?ARTICLE_ID=2908 18&p=1 09]

[86] Ibid.

[87] According to the National Science Foundation, twenty-four states, the Commonwealth of Puerto Rico, and the U.S. Virgin Islands are currently eligible to compete in the NationalScience Foundation's EPSCoR program opportunities. The states are: Alabama, Alaska, Arkansas, Delaware, Hawaii, Idaho, Kansas, Kentucky, Louisiana, Maine, Mississippi Montana, Nebraska, Nevada, New Hampshire, New Mexico, North Dakota, Oklahoma, Rhode Island, South Carolina, South Dakota, Tennessee, Vermont, West Virginia, and Wyoming. [http://www.nsf.gov/od/oia/programs/epscor/eligible.j sp] For more information on EPSCoR, see CRS Report RL30930, U.S. National Science Foundation: Experimental Program to Stimulate Competitive Research (EPSCoR), by Christine M. Matthews.

In: Nanotechnology in the USA: Developments, Policies and Issues ISBN 978-1-60692-800-4
Editor: Carl H. Jennings

Chapter 3

NANOTECHNOLOGY AND ENVIRONMENTAL, HEALTH, AND SAFETY: ISSUES FOR CONSIDERATION*

John F. Sargent

ABSTRACT

Nanotechnology — a term encompassing nanoscale science, engineering, and technology — is focused on understanding, controlling, and exploiting the unique properties of matter that can emerge at scales of one to 100 nanometers. A key issue before Congress regarding nanotechnology is how best to protect human health, safety, and the environment as nanoscale materials and products are researched, developed, manufactured, used, and discarded. While the rapidly emerging field of nanotechnology is believed by many to offer significant economic and societal benefits, some research results have raised concerns about the potential adverse environmental, health, and safety (EHS) implications of nanoscale materials.

Some have described nanotechnology as a two-edged sword. On the one hand, some are concerned that nanoscale particles may enter and accumulate in vital organs, such as the lungs and brains, potentially causing harm or death to humans and animals, and that the diffusion of nanoscale particles in the environment might harm ecosystems. On the other hand, some believe that nanotechnology has the potential to deliver important EHS benefits such as reducing energy consumption, pollution, and greenhouse gas emissions; remediating environmental damage; curing, managing, or preventing diseases; and offering new safety-enhancing materials that are stronger, self-repairing, and able to adapt to provide protection.

Stakeholders generally agree that concerns about potential detrimental effects of nanoscale materials and devices — both real and perceived — must be addressed to protect and improve human health, safety, and the environment; enable accurate and efficient risk assessment, risk management, and cost-benefit trade-offs; foster innovation and public confidence; and ensure that society can enjoy the widespread economic and societal benefits that nanotechnology may offer. Congressionally-mandated reviews of the National Nanotechnology Initiative (NNI) by the National Research Council and the President's Council of Advisors on Science and Technology have concluded that additional research is required to make a rigorous risk assessment of nanoscale materials.

* This is an edited, excerpted and augmented edition of a CRS Report RL34614, dated August 6, 2008.

INTRODUCTION

Nanotechnology — a term encompassing nanoscale science, engineering, and technology — is focused on understanding, controlling, and exploiting the unique properties of matter that can emerge at scales of one to 100 nanometers.[1] These properties are believed by many to offer substantial economic and societal benefits.

A key issue before Congress regarding nanotechnology is how best to protect human health, safety, and the environment as nanoscale materials and products are researched, developed, manufactured, used, and discarded. While the rapidly emerging field of nanotechnology is believed by many to offer significant economic and societal benefits, some research results have raised concerns about the potential environmental, health, and safety (EHS) implications of nanoscale materials. Potential tools the Federal government might use to address these concerns include research and development, regulation, and international engagement.

Some of the properties of nanoscale materials (e.g., small size, high surface area-to-volume ratio) that have given rise to great hopes for beneficial applications have also given rise to concerns about their potential adverse implications for the environment, and human health and safety.[2] There are more than 600 nanotechnology products reportedly commercially available,[3] and with this number of products concerns have been raised about the health and safety of the scientists working with nanoscale materials, workers who manufacture the products, consumers who use the products, and members of the general public who may be exposed to nanoparticles, as well as the environmental impact of nanomanufacturing processes and the use and disposal of nanotechnology products.

Nanoscale particles can result from a variety of different processes. While nanoscale particles can occur naturally (e.g., some particles produced by forest fires, sea spray, volcanoes) and as an incidental by-product of human activities (e.g., some particles contained in welding fumes, diesel exhaust, industrial effluents, cooking smoke), EHS concerns have focused primarily on nanoscale materials that are intentionally designed and produced, often referred to as engineered nanomaterials.

Issues surrounding the potential EHS implications of nanotechnology emerged with the launch in 2000 of the National Nanotechnology Initiative (NNI). The NNI is a multi-agency federal effort to coordinate and expand federal nanotechnology research and development (R&D) efforts. Between FY200 1 and FY2008, the federal government invested $8.4 billion in nanotechnology R&D. Many governments around the world have followed the U.S. lead and established their own national nanotechnology programs. The private sector has invested heavily as well. Global nanotechnology R&D investments — public and private — are estimated to have totaled $12.4 billion in 2006 alone.[4]

Such large investments and intensified efforts to capitalize on these public and private investments have caused some observers (as detailed later in this chapter) to suggest that there is insufficient information about the potential effects nanotechnology products and manufacturing processes may have on human health, safety, and the environment. They assert a variety of uncertainties, including: how nanoscale particles might be transported in air, water, and soil; how they might react with the environment chemically, biologically, or through other processes; how they might be distributed and deposited; and whether they might accumulate in plants or animals.

Others express the view that concerns about nanotechnology EHS implications are often overgeneralized and overstated. Among the arguments they put forth are that nanoscale materials are frequently embedded in other materials as part of the manufacturing process; that some nanotechnology products, such as semiconductors, have nanoscale features but do not contain nanoscale particles; that nanotechnology materials may replace other materials that have significant and known risks; that some nanoscale particles tend to aggregate or agglomerate in the environment into larger particles that no longer have nanoscale dimensions; and that people are regularly exposed to nanoscale particles produced naturally and as incidental byproducts of human activities.

Congressionally-mandated reviews of the NNI by the National Research Council (NRC) and the President's Council of Advisors on Science and Technology (PCAST) have concluded that additional research is required to make a rigorous risk assessment of nanoscale materials. In addition, the NRC warned that, until such information is available, precautionary measures should be taken to protect the health and safety of workers, the public, and the environment.

Nevertheless, most stakeholders agree that these concerns about the potential detrimental effects of nanoscale materials and devices — both real and perceived — must be addressed. Among the issues these stakeholders have identified are characterizing the toxicity of nanoscale materials; developing methods for assessing and managing the risks of these materials; and understanding how these materials move in, and interact with, the environment.

This chapter identifies the potential environmental, health, and safety opportunities and challenges of nanotechnology; explains the importance of addressing nanotechnology EHS concerns; identifies and discusses nanotechnology EHS issues; and summarizes options for Congressional action, including the nanotechnology EHS-related provisions of selected legislation. The chapter also includes two appendices. Appendix A provides an overview of selected nanotechnology EHS activities of federal regulatory agencies. Appendix B provides an overview of selected EHS-related international engagement efforts of NNI agencies.

For more information on the NNI see, CRS Report RL3440 1, The National Nanotechnology Initiative: Overview, Reauthorization, and Appropriations Issues, by John F. Sargent. For additional information on nanotechnology-related regulatory challenges, see CRS Report RL34332, Engineered Nanoscale Materials and Derivative Products: Regulatory Challenges, by Linda-Jo Schierow.

Opportunities and Challenges

Historically, many new technologies have delivered general societal benefits while presenting EHS challenges. For example, automobiles increased personal mobility and provided faster, less expensive transportation of goods, but soon became a leading cause of accidental deaths and injuries, as well as a source of emissions that can damage air quality and may contribute to global climate change. Similarly, genetically-modified (GM) plants have traits such as greater resistance to pests, pesticides, or cold temperatures that contribute to higher crop yields, while critics argue some GM foods contribute to food allergies and antibiotic resistance.[5]

Like other new technologies, nanotechnology offers potential economic and societal benefits, and presents potential EHS challenges as well. Nanotechnology advocates assert, however, that nanotechnology provides the opportunity to reduce or eliminate known risks by engineering around them. Proponents maintain that nanotechnology also offers the potential for significant EHS benefits, including:

- reducing energy consumption, pollution, and greenhouse gas emissions;
- cleaner, more efficient industrial processes;
- remediating environmental damage;
- curing, managing, or preventing deadly diseases; and
- offering new materials that protect against impacts, self-repair to prevent catastrophic failure, or change in ways that protect or aid soldiers on the battlefield.

For example, nanoscale materials show promise for detecting, preventing, and removing pollutants. According to the Environmental Protection Agency (EPA):

> nanoscale cerium oxide has been developed to decrease diesel engine emissions; iron nanoparticles can remove contaminants from soil and ground water; and nano-sized sensors hold promise for improved detection and tracking of contaminants.[6]

In the area of human health, scientists assert nanotechnology has the potential for improving disease diagnostics, sensing, monitoring, assessment, and treatment. In particular, the National Cancer Institute (NCI) views nanotechnology as likely to provide revolutionary tools to extend and improve lives. In July 2004, NCI launched a five-year, $145 million initiative focused on applying nanotechnology to the prevention, detection, and treatment of cancer and amelioration of its symptoms. At the initiative's launch, then-NCI Director Andrew von Eschenbach identified nanotechnology as a key component of the agency's strategy for ending death and suffering from cancer by 2015 (see text box, "Potential Nanotechnology Cancer Applications").[7]

Some characteristics of nanoscale particles could produce both positive and negative consequences. According to E. Clayton Teague, director of the National Nanotechnology Coordination Office (NNCO),

the unique properties of these [nanotechnology] materials are a double-edged sword: they can be tailored for beneficial properties, but also have unknown consequences, such as new toxicological and environmental effects.[8]

The following examples illustrate how the same nanotechnology material may be both potentially beneficial and potentially harmful:

- Nanoscale silver is highly effective as an antibacterial agent in wound dressings, clothing, and washing machines, but some have expressed concerns that widespread dispersion of nanoscale silver in the environment could kill microbes that are vital to waste water treatment plants and to ecosystems. Some beneficial bacteria, for example, break down organic matter, remove nitrogen from water, aid in animal digestion, protect against fungal infestations, and even aid some animals in defense against predators.[9]

- Some nanoscale particles may have the potential to penetrate the blood-brain barrier, a structure that protects the brain from harmful substances in the blood but also hinders the delivery of therapeutic agents. The characteristics of certain nanoscale materials may allow pharmaceuticals to be developed to purposefully and beneficially cross this barrier and deliver medicine directly to the brain to treat, for example, a brain tumor.[10] Some critics are concerned, however, that nanoscale particles might unintentionally pass through the blood-brain barrier causing harm to humans and animals.[11]
- Certain nanoscale materials are highly chemically reactive due to their high surface-to-volume ratio.[12] This is a property that might be positively exploited in catalysis, treatment of groundwater contamination, and site remediation. This property also is being explored for use in protective masks and clothing as a defense against chemical and biological agents. However, some research results indicate that the reactivity of some nanoparticles potentially can result in cell damage in animals.[13]
- Carbon nanotubes (CNTs) have potential uses in a wide range of applications (e.g., materials, batteries, memory devices, electronic displays, transparent conductors, sensors, medical imaging). However, some scientists have expressed concerns that some CNTs exhibit properties similar to asbestos fibers, and might become lodged in organs (e.g., lungs), harming humans and animals.[14]

Potential Nanotechnology Cancer Applications

The NCI Cancer Nanotechnology Plan asserts that nanotechnology can serve as an enabling technology for a variety of cancer-related applications:

- imaging agents and diagnostics that allow clinicians to detect cancer in its earliest, most easily treatable, pre-symptomatic stage;
- systems that provide real-time assessments of therapeutic and surgical efficacy;
- multifunctional, targeted devices capable of bypassing biological barriers to deliver therapeutic agents at high local concentrations directly to cancer cells and tissues that play a critical role in the growth and metastasis of cancer;
- agents capable of monitoring predictive molecular changes and preventing precancerous cells from becoming malignant;
- surveillance systems that detect mutations that may trigger the cancer process and genetic markers that indicate a predisposition for cancer;
- novel methods for managing the symptoms of cancer that adversely impact quality of life; and
- research tools that enable investigators to quickly identify new targets for clinical development and predict drug resistance.

Source: Cancer Nanotechnology Plan: A Strategic Initiative to Transform Clinical Oncology and Basic Research Through the Directed Application of Nanotechnology, National Cancer Institute, National Institutes of Health, Department of Health and Human Services, July 2004.

EHS Concerns about Carbon Nanotubes and other Fullerenes

Much of the public dialogue about potential risks associated with nanotechnology has focused on carbon nanotubes (CNTs) and other fullerenes (molecules formed entirely of carbon atoms in the form of a hollow sphere, ellipsoid, or tube) since they are currently being manufactured and are among the most promising nanomaterials. These concerns have been amplified by some research on the effects of CNTs on animals and on animal and human cells. For example, researchers have reported that carbon nanotubes inserted into the trachea of mice can cause lung tissue damage;[a] that buckyballs (spherical fullerenes) caused brain damage in fish;[b] and that buckyballs can accumulate within cells and potentially cause DNA damage.[c]

There are scientists who have argued that experiments indicating CNT/fullerene toxicity are not conclusive. They suggest that toxicity reported by researchers may have resulted from uncharacterized contaminants in the samples resulting from the synthesis, purification, and post-processing methods used in the manufacture of CNTs. Thus, they assert, the experiments could be measuring the toxicity of non-nanoscale materials and, therefore, unfairly indicting nanoscale materials. They also contend that such non-nanoscale contaminants, if identified as toxic, potentially could be eliminated or controlled in the manufacturing process. The issue of contaminants is often cited by advocates for improved standards, reference materials, sensors, instrumentation, and other technologies for the characterization of nanoscale materials.

Some experiments have produced results that indicate CNTs/fullerenes are non-toxic. Research on single-walled carbon nanotubes (SWCNTs) by the Institute of Toxicology and Genetics in Karlsruhe, Germany, reported that, in three of four different types of tests conducted, SWCNTs did not show toxicity. In the fourth test, which appeared to indicate SWCNT toxicity, the researchers concluded that the results were a "false positive" and explained how the SWCNTs interacted with the materials in the assay to produce a misleading result. These researchers concluded that this result points to the need for careful selection of assays and the need for the establishment of standards for toxicity testing of CNTs and other nanomaterials.[d]

Work at Rice University's Center for Biological and Environmental Nanotechnology conducted in 2005 found cell toxicity of CNTs to be low, and that it could be reduced further through simple chemical changes to the surface.e Earlier research demonstrated that similar surface modifications of buckyballs reduced their toxicity. Nanotechnology may offer the potential to engineer around known and potential hazards by changing the size, molecular construction, or other property of a nanoscale material to make it safe or less hazardous. Experts advise that the potential to do so will require a thorough understanding of the properties of the various nanoparticles and their effects on humans and other organisms

a Lam, C.W., James, J.T., McCluskey, R., and Hunter, R.L. "Pulmonary toxicity of single-wall carbon nanotubes in mice 7 and 90 days after intratracheal instillation," September 2003. [http://www.ncbi.nlm.nih.gov/sites/entrez?cmd= Retrieve&db= PubMed&list_uids= 14514958]

b Oberdörster, Eva. "Manufactured Nanomaterials (Fullerenes, C60) Induce Oxidative Stress in the Brain of Juvenile Largemouth Bass," April 2004. [http://www.pubmedcentral.nih.gov/articlerender.fcgi?tool=pmcentrez&artid= 1247377] c Magrez, A., Kasas, S., Salicio, V., Pasquier, N., Seo, J.W., Celio, M., Catsicas, S., Schwaller, B., and Forro, L. "Cellular Toxicity of Carbon-Based Nanomaterials," Nano Letters, 6(6):1 121-1125, American Chemical Society, May 2006. [http://pubs.acs.org/cgi-bin/abstract.cgi/nalefd/ 2006/6/i06 /abs/nl060162e.html]

d Wörle-Knirsch, J.M., Pulskamp, K., and Krug, H. F. "Oops They Did It Again! Carbon Nanotubes Hoax Scientists in Viability Assays," American Chemical Society, Nano Letters, Vol.6, April 2006. "Modifications render carbon nanotubes nontoxic," press release, Rice University, October 2005.

IMPORTANCE OF ADDRESSING EHS ISSUES

Nanotechnology covers a wide swath of scientific fields, engineering disciplines, and technological applications. Sufficient knowledge has been developed about the useful properties of certain nanomaterials, how they can be manufactured, and how they can be applied in useful ways to enable commercial product development. In other areas of nanotechnology, fundamental research on nanoscale phenomena and processes is under way that may lead to greater understanding and beneficial applications in the years ahead. In general, however, nanotechnology is still an emerging field and there is a dearth of information about how nanoscale particles and devices might adversely affect human health, safety, and the environment. Accordingly, there is widespread agreement on the need for more research to better understand such implications.

In reviews of the NNI, [15] both the National Research Council and the President's Council of Advisors on Science and Technology (PCAST) concluded that assessment of potential nanotechnology EHS risks is not possible due to the absence of information and tools. According to the NRC,

> it is not yet possible to make a rigorous assessment of the level of risk posed by [engineered nanomaterials]. Further risk assessment protocols have to be developed, and more research is required to enable assessment of potential EHS risks from nanomaterials.[16]

Similarly, PCAST concluded that

> it is premature to rigorously assess the levels of risk posed by engineered nanomaterials. Adequate tools are being developed but are not yet in place.[17]

Leaders of the NNI have argued strongly that to achieve the economic, societal, and EHS benefits of nanotechnology the nation must concurrently address its potential adverse effects. According to then-Under Secretary of Commerce for Technology Phillip J. Bond, a leading Administration advocate for the NNI, Addressing societal and ethical issues is the right thing to do and the necessary thing to do. It is the right thing to do because as ethically responsible leaders we must ensure that technology advances human well-being and does not detract from it. It is the necessary thing to do because it is essential for speeding technology adoption, broadening the economic and societal benefits, and accelerating and increasing our return on investment.[18]

This is a view shared by many in the business community. A 2006 survey of business leaders in the field of nanotechnology indicated that nearly two-thirds believe that "the risks to the public, the workforce, and the environment due to exposure to nano particles are 'not known,'" and 97% believe that it is very important or somewhat important for the government to address potential health effects and environmental risks that may be associated with nanotechnology.[19]

The Project on Emerging Nanotechnologies (PEN) has warned that bad practices in nanotechnology research or production may result in a nanotechnology accident that would

chill investment, galvanize public opposition, and generally lead to a lot of hand wringing on the part of governments who are betting large sums of money on the nanotech revolution.[20]

Successfully addressing EHS issues is seen as vital for those potentially exposed to nanoscale materials (e.g., consumers, researchers, manufacturing workers, the general public), businesses, and investors for a variety of reasons:

- protecting and improving human health, safety, and the environment;
- enabling accurate and efficient risk assessments, risk management, and cost-benefit trade-offs;
- ensuring public confidence in the safety of nanotechnology research, engineering, manufacturing, and use;
- preventing a problem in one application area of nanotechnology from having negative consequences for the use of nanotechnology in unrelated application areas due to public fears, legislative interventions, or an overly-broad regulatory response; and
- ensuring that society can enjoy the widespread economic and societal benefits that nanotechnology is believed by many to offer.

In addition, the U.S. regulatory environment for nanotechnology could be an enabler for innovation and contribute to a strong, sustainable economy by creating predictability, accurately assessing risks and benefits, and fostering the swift movement of safe products into the market. Such an environment is likely to favor nanotechnology-related investments and innovative activities in the United States by domestic and foreign stakeholders, as opposed to nations where such regulatory conditions do not exist.

Conversely, if the U.S. regulatory environment is not handled effectively (i.e., if it lacks predictability, if regulatory approaches do not accurately assess risks and benefits, or if approval processes are too long or expensive) it could prove a major impediment to innovation, economic growth, and job creation, as well as posing a potential threat to health, safety, and the environment. In such a regulatory environment, investment capital may be driven away from nanotechnology, potentially beneficial products may not be developed, safe products may be denied regulatory approval, or unsafe products may be allowed to enter the market.

Alternatively, nanotechnology investments, research, and production may be driven to other nations with preferable regulatory environments. On the one hand, such a regulatory system might be more desirable to investors and companies because it is more predictable, more efficient, and less costly. In such a case, the United States might miss out on nanotechnology' s potential economic benefits. On the other hand, if other nations' regulatory systems are more attractive to investors and producers because those systems under-regulate or do not regulate at all, then nanotechnology research, development, and production could present increased EHS risks worldwide.

Selected Issues for Consideration

Given the widespread agreement that nanotechnology EHS concerns must be addressed, discourse on how best to do so has focused on three main issues:

- federal investment in EHS research;
- federal regulation; and
- international engagement.

These issues are closely interrelated. For example, reliable EHS research is required by regulatory bodies to determine whether and how to regulate nanotechnology products. Since all nations face the same fundamental health, safety, and environmental issues, international coordination on EHS research could help accelerate development of a common body of knowledge through the sharing of results and reduction in redundant research. This shared knowledge could, in turn, inform regulatory decision making and perhaps improve the consistency of regulations among nations. Regulations, standards, and enforcement might need to be coordinated worldwide to protect workers and consumers as intermediate and final products are frequently produced along global supply chains and sold in industrial and commercial markets around the world. In addition, one nation's policies governing nanotechnology production, use, and disposal may have implications for nearby nations and, perhaps, for all nations.

Federal Investment in EHS Research

Current Funding Level

There is not a single, centralized source of EHS research funds that is allocated to individual agencies. Agency nanotechnology budgets are developed internally as part of each agency's overall budget development process. These budgets are subjected to review, revision, and approval by the Office of Management and Budget (OMB) and become part of the President's annual budget submission to Congress. The NNI budget — and the EHS component — is then calculated by aggregating the nanotechnology components of the appropriations provided by Congress to each federal agency. While there is some coordination of EHS-research budget requests through the Nanotechnology Environmental and Health Implications (NEHI) working group[21] and in OMB's budget development process, the decision process that establishes overall funding for nanotechnology EHS research is highly decentralized.

In FY2008, NNI funding for EHS implications research [22] is $58.6 million, approximately 3.9% of the total NNI budget of $1.49 billion. This represents an increase over the FY2007 EHS research level of $48.3 million (3.4% of the total NNI budget), and the FY2006 level of $37.7 million (2.8%), both in dollars and in share of total NNI funding.

President Bush has requested $76.4 million (5.0%) for EHS research in FY2009. NNI EHS research funding for FY2006 through FY2008, and the request for FY2009, is provided in Table 1.

Table 1. NNI Environmental, Health, and Safety Research Funding, FY2006-2008

	EHS research, in current dollars	EHS research's share of total NNI budget
FY2006 (actual)	$ 37.7 million	2.8%
FY2007 (actual)	48.3 million	3.4%
FY2008 (estimated)	58.6 million	3.9%
FY2009 (requested)	76.4 million	5.0%

Sources: "The National Nanotechnology Initiative: Research and Development Leading to a Revolution in Technology and Industry, Supplement to the President's FY2008 Budget," NSET Subcommittee, NSTC, OSTP, The White House, July 2007; "National Nanotechnology Initiative: FY2009 Budget and Highlights," NSET Subcommittee, NSTC, The White House, February 2008. [http://www.nano.gov/NNI_FY09_budget_summary.pdf].

NNI officials assert that the initiative also conducts EHS research as a part of its other research activities, but that these EHS investments are not easily quantified and thus are not reflected in the NNI' s reported figure for EHS funding. PCAST agreed with this assertion in its 2008 assessment, arguing that

> In many instances, nanotechnology EHS research cannot be separated from the particular application(s) research and from the context for which a specific nanomaterial is intended. Such division is unproductive and neglects the whole benefit of research. Consequently, [PCAST] expects that a substantial fraction of nanotechnology research related to EHS will continue to take place under the auspices of agencies that fund applications R&D and may not be uniquely or exclusively identified as nanotechnology EHS research.... Furthermore, detailed reporting on the degree of relevance to EHS of such research is not necessarily critical to (and may actual hinder) overall prioritization and coordination.[23]

In 2007, OMB issued a one-time request to all NNI research agencies to report funding data on research related to the five categories identified in the NSET document, Prioritization of Environmental, Health, and Safety Research Needs for Engineered Nanoscale Materials.[24] Totals for EHS implications research spending identified in each of the five categories is shown below in Table 2. Preliminary analysis of this data by the NEHI working group indicated that $67 million was spent on EHS research in FY2006, in contrast to the reported figure of $37.7.

Table 2. FY2006 NNI Funding for EHS Research by Research Needs Categories

Category	Estimated Funding
Instrumentation, Metrology, and Analytical Methods	$27 million
Nanomaterials and Human Health	$24 million
Nanomaterials and the Environment	$13 million
Health and Environmental Exposure Assessment	$ 1 million
Risk Management Methods	$ 3 million
TOTAL	$67 million

Note: Numbers may not add due to rounding.

Source: Teague, E. Clayton, director, National Nanotechnology Coordination Office. Testimony before the Subcommittee on Research and Science Education, Committee on Science and Technology, U.S. House of Representatives. Hearing on "Research on Environmental and Safety Impacts of Nanotechnology: Current Status of Planning and Implementation under the National Nanotechnology Initiative." 110th Cong., 1st Sess., October 31, 2007.

Critics (as detailed in the following section) assert that the current level of federal nanotechnology EHS research is too low and represents too small a share of the overall NNI budget. These critics argue that the current allocation of NNI funding may produce a flood of products for which there is inadequate information to assess and manage their EHS risks.

However, executive branch officials stress that the United States leads the world in EHS funding and, by inference, that the current funding level is adequate. White House Office of Science and Technology Policy (OSTP) director John Marburger asserted that the United States

> leads the world not only in spending for nanotechnology development, but also, by an even larger margin, in its investment in research to understand the potential health and safety issues.[25]

Similarly, NNCO director E. Clayton Teague asserted U.S. leadership in nanotechnology EHS research:

> During fiscal years 2005 through 2008, it is estimated that NNI agencies will have invested nearly $180 million in research whose primary purpose is to address the EHS implications of nanomaterials. With these investments, the United States leads all other countries by a wide margin in support of such research.[26]

Dr. Teague maintains that EHS research has been a top priority of the Administration and the NNI, citing, as an example, the annual R&D budget guidance memorandum sent by the directors of OMB and OSTP to departments and agencies. This memorandum identifies Administration priorities and is intended to help guide agency budget development for the following fiscal year. The OMB/OSTP memorandum to guide FY2006 agency budget development stated that

> In order to ensure that nanotechnology research leads to the responsible development of beneficial applications, agencies also should support research on the various societal implications of the nascent technology. In particular, agencies should place a high priority on research on human health and environmental issues related to nanotechnology and develop, where applicable, cross-agency approaches to the funding and execution of this research.[27]

The OMB/OSTP memorandum has included similar language in each succeeding year.

In their reviews of the NNI, both the NRC and PCAST concluded that federal EHS research funding should be expanded. According to the NRC assessment,

> To help ensure the responsible development of nanotechnology ... research on the environmental, health, and safety effects of nanotechnology [should] be expanded.[28]

PCAST acknowledged potential EHS risks in its first review of the NNI but found the federal government was "directing appropriate attention" and "adequate resources" to EHS research. In its second assessment, PCAST termed the current federal investment level in EHS "appropriate," but added that

> expanded EHS research, broad-based protocol development, and particularly standardization are necessary .... the funding level for EHS [should] continue to grow consistent with the needs identified in the NNI research strategy for nanotechnology EHS as well as the available capacity for quality research.[29]

Alternative Approaches

Various alternatives have been suggested for addressing the perceived shortcoming in EHS funding. One recommendation is requiring a fixed percentage of the NNI's total funding be devoted to EHS research. A figure of 10% has been proposed for this purpose by organizations such as the NanoBusiness Alliance and the Project on Emerging Nanotechnologies. If this proposal had been in effect in FY2008, the NNI would have been required to spend \$149 million on EHS research, more than twice as much as the NSET-reported level of \$58.6 million. In testimony before the House Committee on Science and Technology, Sean Murdock, executive director of the NanoBusiness Alliance, agreed with the level of funding represented by the 10% figure but argued the need for cross- agency flexibility in achieving it:

> The NanoBusiness Alliance believes that environmental, health, and safety research should be fully funded and based on a clear, carefully-constructed research strategy. While we believe that 10 percent of the total funding for nanotechnology research and development is a reasonable estimate of the resources that will be required to execute the strategic plan, we also believe that actual resource levels should be driven by the strategic plan as they will vary significantly across agencies.[30]

Others have suggested a different approach, proposing fixed dollar amounts or minimum levels. For example, the Environmental Defense Fund has called for $100 million or more in federal nanotechnology EHS research funding.[31]

In its 2008 assessment, PCAST disagreed with both approaches:

> growing research in nanotechnology EHS must be strategic, guided by ... a comprehensive set of scientifically determined priorities and needs rather than arbitrary percentages or funding figures.[32]

By establishing a 10 percent requirement (or setting a figure of $100 million for total EHS funding), the United States could greatly accelerate the growth in EHS research spending. In testimony before Congress in 2007, PCAST co-chair Floyd Kvamme warned against such a rapid increase:

> In general, increasing funding too rapidly does not lead to equivalent increases in high quality research. It is crucial to note that EHS research also depends on advances in non-EHS areas, such as instrumentation development and basic research on nanomaterials.[33]

Some non-governmental organizations (NGOs) have advocated for a more restrained approach to nanotechnology research and development. They assert that the federal government is pushing ahead too quickly in developing nanotechnology and encouraging its commercialization and use without sufficient knowledge and understanding of EHS implications and how they might be mitigated.[34] They argue that the very characteristics that make nanotechnology promising also present significant potential risks to human health and safety and the environment. Some of these groups argue for application of the "precautionary principle,"35 which holds that regulatory action may be required to control potentially hazardous substances even before a causal link has been established by scientific evidence.36 In 2006, Friends of the Earth warned that

> The early warning signs surrounding nanotoxicity are serious and warrant a precautionary approach to the commercialization of all products containing nanomaterials.... there should be a moratorium on the further commercial release of sunscreens, cosmetics and personal care products that contain engineered nanomaterials, and the withdrawal of such products currently on the market, until adequate public, peer-reviewed safety studies have been completed, and adequate regulations have been put in place...[37]

The Action Group on Erosion, Technology, and Concentration (ETC Group) has called for a moratorium on the conduct of nanotechnology R&D and use of commercial products incorporating man-made nanoparticles:

> Given the concerns raised over nanoparticle contamination in living organisms, Heads of State ... should declare an immediate moratorium on commercial production of new nanomaterials and launch a transparent global process for evaluating the socio-economic, health and environmental implications of the technology.[38]

In 2003, the ETC Group expanded the breadth of its proposed moratorium:

> In the absence of toxicology studies, ETC Group believes that governments must also urgently consider extending the moratorium to products that place consumers in direct contact with synthetic nanoparticles through their skin, lungs or digestive systems.[39]

In contrast to these views, a report prepared by the NSET Subcommittee concluded that conducting EHS research in parallel with the development of nanomaterials and their applications will help to ensure the full, safe, and responsible realization of the promise of nanotechnology.[40]

In 2003, then-Under Secretary of Commerce for Technology Phillip J. Bond addressed calls for a moratorium or slowdown in nanotechnology R&D, casting the issue in ethical terms:

> Those who would have us stop in our tracks argue that it is the only ethical choice. I disagree. In fact, I believe a halt, or even a slowdown, would be the most unethical of choices.... Given the promise of nanotechnology, how can our attempt to harness its power at the earliest opportunity — to alleviate so many of our earthly ills — be anything other than ethical? Conversely, how can a choice not to attempt to harness its power be anything other than unethical?[41]

Management of Federal EHS Research

In order to manage the Federal EHS portfolio, policymakers will need to establish research priorities. In this regard, the NRC recommended that

> Assessing the effects of engineered nanomaterials on public health and the environment requires that the research conducted be well defined and reproducible and that effective methods be developed and applied to (1) estimate the exposure of humans, wildlife, and other ecological receptors to source material; (2) assess effects on human health and ecosystems of both occupational and environmental exposure; and (3) characterize, assess, and manage the risks associated with exposure.[42]

In 2005, PCAST concluded that EHS research should give highest priority to workplace exposure. PCAST noted

> the greatest likelihood of exposure to nanomaterials is during manufacture, and therefore [we] agree with the prioritization of research on potential hazards from workplace exposure.[43]

Several years later, in its 2008 assessment, PCAST reiterated this point stating, "the greatest risk of exposure to nanomaterials at present is to workers who manufacture or handle such material," but also acknowledged a broader range of risks:

> environmental, health, and safety risks in a wide range of settings must be identified and the necessary research performed so that real risks can be appropriately addressed.[44]

Some stakeholders have asserted that a comprehensive approach to federal EHS research has been hampered by the lack of an NNI roadmap for these efforts.45 In general, these stakeholders seek a multi-year roadmap with specific milestones, metrics, and funding levels. Such a roadmap, they assert, would contribute to a more coordinated approach among agencies and between the executive branch and Congress on the magnitude, timing, prioritization, and management of federal EHS research.

NNI officials argue that the NSET Subcommittee, the coordinating body for the NNI, has developed an EHS research strategy and articulated it in three reports (see text box, "NNI EHSfocused Reports"), though they acknowledge that these documents do not constitute a roadmap. At an October 2007 hearing of the House Subcommittee on Research and Education,[46] some Members of Congress expressed concerns about the time required by the National Nanotechnology Coordination Office to produce a prioritized, detailed implementation plan for NNI EHS research. While acknowledging the challenges faced by the NNCO in developing consensus among the 25 NNI agencies, some Members suggested that these challenges were emblematic of the need for a more top-down approach to EHS research.

Opposition to an EHS roadmap stems primarily from doubts of the practicality and efficacy of such an approach. Some argue that it is unlikely that OMB would commit to a multi-year, multi-agency roadmap accompanied by specific funding levels. Such an approach would depart from the current executive branch annual budget development process and reduce OMB's flexibility in future years. In addition, agencies often have to respond to new requirements based on emergent circumstances, Congressional direction, or other factors. Agency funding is often redirected from planned efforts to new, often imminent, priorities. The need for such redirection of funding could impede the achievement of roadmap milestones and metrics or, conversely, impede the movement of funding to new priorities.

To overcome the obstacles associated with the development of a roadmap by the agencies, some have suggested the National Academies produce such a roadmap. Some assert that this approach worked well with respect to the development of a federal research roadmap to reduce EHS uncertainties associated with airborne particulate matter. Others argue that the particulate matter effort focused only a narrow field and covered research conducted by only a single agency (EPA); in contrast, nanotechnology spans a broad range of materials and applications across many fields, and requires EHS research efforts by several agencies.

In February 2007, 19 environmental and business organizations, large and small companies, and research organizations signed a letter to the Senate Appropriations Subcommittee on Interior, Environment, and Related Agencies requesting $1 million be appropriated for the development of a federal roadmap and research strategy. The letter recommended that this work be done by the National Institute of Environmental Health Sciences (NIEHS).[47]

The Senate Appropriations Committee report (S.Rept. 110-91) accompanying the Department of the Interior, Environment, and Related Agencies Appropriations Act, 2008[48] urged the Environmental Protection Agency (EPA) to

> contract or enter into a cooperative agreement with the National Academy of Sciences' Board on Environmental Studies and Toxicology within 90 days of enactment to develop and monitor implementation of a comprehensive, prioritized research roadmap for all Federal agencies on environmental, health and safety issues for nanotechnology.[49]

NNI EHS-focused Reports

Environmental Health and Safety Research Needs for Engineered Nanoscale Materials, published in September 2006, identified the research and information needed to enable sound risk assessment and risk management decision making with respect to nanoscale materials and products that incorporate them.

Prioritization of Environmental, Health, and Safety Research Needs for Engineered Nanoscale Materials, published in August 2007, identified five broad categories of EHS research and information needs, and five specific research areas in each category.

The National Nanotechnology Initiative: Strategy for Nanotechnologyrelated Environmental, Health, and Safety Research, published in February 2008, defined the NNI' s strategy for addressing priority research on EHS aspects of nanomaterials. The document reviewed current agency research using the taxonomy developed in the second report; identified research gaps; and articulated a framework for prioritizing research, implementing the strategy, and coordinating agency efforts.

A Cooperative Approach to Addressing EHS Concerns

Some organizations have taken a cooperative approach to promote EHS research. For example, the Environmental Defense Fund, an environmental advocacy group, partnered with the American Chemistry Council, a trade group, to issue a Joint Statement of Principles in June 2005 that recognizes the "significant societal and sustainable development benefits" expected from nanotechnology, while calling for a multi-stakeholder dialogue to achieve the timely development of nanomaterials "in a way that minimizes potential risks to human health and the environment." The statement also called for increased federal investments in EHS research and

> development of an international effort to standardize testing protocols, hazard and exposure assessment approaches, and nomenclature and terminology ... to maximize resources and minimize inconsistent regulation of nanomaterials.[a]

There is general agreement among stakeholders that these activities can contribute to creating an environment where research results can be reliably shared and compared, to protecting human health and safety, and to creating a common language about nanotechnology that increases clarity in the sharing of ideas and information. However international standardization efforts are often time- and resource-consuming, and can divert resources from more pressing needs. In addition, such efforts can be used by nations and other organizations for competitive advantage (e.g., by securing the adoption of a favorable standard, slowing others' progress).

In June 2007, the Environmental Defense Fund and DuPont issued a Nano Risk Framework "to assist with the responsible development and use of nanotechnology and to help inform global dialogue on its potential risks."[b] The framework is a six-step process to identify, address, and manage potential risks: (1) describe the material and the intended application; (2) profile the material's lifecycle in the application; (3) evaluate associated risks;

(4) assess risk management options; (5) decide on and document actions; and, (6) regularly review new information and adapt actions accordingly.[c]

Environmental Defense and American Chemistry Council Nanotechnology Panel: Joint Statement of Principles, Comments on EPA's Notice of a Public Meeting on Nanoscale Materials, June 23, 2005.

"DuPont and Environmental Defense Launch Comprehensive Tool for Evaluating and Addressing Potential Risks of Nanoscale Materials," press release, E. I. du Pont de Nemours and Company, June 21, 2007. [http://vocuspr.vocus.com/Vocu sPR3 0/Newsroom/Query.aspx?SiteName=
DupontNew&Entity=PRAsset&SF_PRAsset_PRAssetID_EQ=106677&XSL=PressRelease&
Cache=False]

Nanorisk Framework, Environmental Defense-DuPont Nano Partnership, June 2007.

The process used to develop research priorities and the federal EHS budget has also raised management concerns. As discussed earlier, the federal nanotechnology EHS research portfolio results from research funding requests made by individual agencies pursuing their missions and by decisions made in the Congressional appropriations process. Informal research coordination among EHS funding agencies occurs through the NEHI working group and more formally through the OMB budget development process. Some proponents for an integrated federal EHS research effort have called for a more top-down approach. The Woodrow Wilson Center's Project on Emerging Nanotechnologies (PEN) has been a leading advocate on this issue. PEN's chief science advisor, Andrew Maynard, asserted that

> to realize nanotechnology' s benefits ... the federal government needs a master plan for identifying and reducing potential risks. This plan should include a top-down risk research strategy, dedicated and sufficient funding to do the job, and the mechanisms to ensure that resources are used effectively.[50]

PEN has recommended increasing the authorities of the NEHI working group to empower it to develop and implement the top-down research plan, a minimum of $100 million over two years to fund the research, and a full-time director to support the NEHI working group.

Responding to the PEN recommendation, E. Clayton Teague, director of the NNCO, testified before Congress that there was a consensus among NNI agencies that a centralized office with budgetary authority to oversee the NNI' s EHS research program would have significant detrimental effects. According to Dr. Teague,

No one agency or centralized organization would have the breadth of scientific expertise and knowledge of regulatory authorities and needs currently represented by the 20 agencies participating in the NEHI working group.

Creation of a new central authority would undermine the existing successful interagency coordination.

Moving the management of all nanotechnology EHS research into a single office would likely decouple such research from related efforts within NNI agencies and from the knowledge base in the agencies that is currently networked into the NNI' s EHS research effort.

Creating a separate office would, on the one hand, give mission agencies a disincentive for doing nanotechnology-related EHS research. They would reasonably assume that another agency is responsible, and they therefore could redirect their limited resources to address other priorities. A likely result could be that the level of research would actually decrease. Conversely, creating a separate office could lead to duplicative work being funded, thereby wasting tax dollars and not optimizing progress.[51]

Dr. Maynard counters that "it should be possible to develop a functional structure that enables agencies to work within a broader plan." According to Maynard, while a centralized office is not necessary,

> top-down leadership with authority and the ability to ensure resources get towhere they are needed is necessary.... [Such] leadership does not take away fromagencies' expertise and missions, but rather empowers agencies to do the bestthey can, while coordinating and partnering as effectively as possible with each other.[52]

Project on Emerging Nanotechnologies Recommendations

The Project on Emerging Nanotechnologies (PEN), a joint venture of the congressionally-chartered Woodrow Wilson Center for International Scholars and the Pew Charitable Trusts, has produced inventories of both nanotechnology-based products and government-funded EHS research. PEN has asserted the need for more EHS research, more aggressive oversight, and a more centralized federal government approach to funding EHS research.

In addition, PEN contends that the increasing complexity of systems incorporating nanoparticles with multiple functions will make the behaviors more complex and difficult to predict. To minimize the likelihood of a nanotechnology accident, PEN made the following recommendations:

- Creating a Nano Safety Reporting System where people working with nanotechnology can anonymously report safety issues and concerns. PEN states that the information gleaned from this system could be used to inform the design of educational materials, better structure technical assistance programs, and provide an early indicator of emerging safety issues.
- Creating technologies that provide an early-warning system to allow for risk to be assessed early in research efforts. Such a technology might enable low-cost, fast-screening for novel properties that would allow for risk assessment integrated and concurrent with the R&D process.
- Pushing information out to small businesses, start-ups, and laboratories that, due to their size and resources, are unlikely to be able to devote significant resources to EHS issues. PEN states that existing assistance programs could be used to deliver this information, as well as the development of peer-to-peer mentoring programs within industrial supply chains.
- Application of lessons learned in other technology areas to make nanotechnology more inherently safe, using strategies such as multiple levels of protection, learning

from failures, not oversimplifying the complex, awareness of operations, and building in resilience to prevent cascading of errors.

Source: Rejeski, David, director, Project on Emerging Nanotechnologies. "Nanotech Safety 101 or How to Avoid the Next Little Accident," paper, Workshop on Disaster Prevention, Harvard University, April 27, 2006.

Unique Properties Emerge at the Nanoscale

Scientists have discovered that elements and materials with the same chemistry can exhibit fundamentally different properties at the nanoscale. For example, platinum, which exhibits no magnetism in its bulk form, shows significant magnetic properties in nanoscale clusters of 13 atoms. The optical properties of gold also can change with particle size. At 10 nanometers, gold particles absorb green light and appear red, not gold.

Not only can nanoscale particles differ in properties from bulk material with the same chemical composition, they may also differ from other nanoscale materials with the same chemical composition. For example, the melting point of an element — which was believed to be constant regardless of the element's particle size — can change with particle size. Nanotechnology research has demonstrated that the melting temperature of gold decreases when the particle's radius drops below 10 nanometers (from a melting temperature of approximately 1,000oC at 10 nanometers to approximately 500oC at 2 nanometers).

Source: Roduner, Emil. "NanoscopicMaterials: Size-Dependent Phenomena,"University of Stuttgart, Germany, August 2006.

Federal Regulation

Some have raised concerns about whether current laws, regulations, and authorities are adequate to protect human health, safety, and the environment from potential adverse implications of nanotechnology. Several factors may affect the ability of the regulatory system to keep pace with advances in technology, both broadly and specifically with respect to nanotechnology.

Broadly, market forces have increased the pace of global innovation, challenging institutions' ability to identify and cope with the societal implications of rapid change. Speed-to-market has become a driving factor in competition for many industries as a result of the entry of new and nimble competitors in the global marketplace, increased public and private investments in R&D, global models of innovation, increased flows of scientific and technical knowledge, and greater numbers of scientists and engineers around the world. In addition, growing global markets enable companies to recoup their investments faster and enable earlier investments in subsequent generations of technology, further accelerating the pace of innovation. The increased pace, scope, and complexity of technological innovation may pose challenges to the existing regulatory system. While these factors may affect a broad range of technologies, nanotechnology may be especially affected due to the rapid growth in public and private R&D investments in the field since the year 2000 and the potential for nanomaterials to be used in a wide array of products.

Nanotechnology also may pose unique challenges to the regulatory system. For example, historically, regulatory agencies have defined a chemical by its chemical composition, usually without regard to its particle size. In contrast, the essence of nanotechnology is that a material may exhibit different properties at the nanoscale than it does at a bulk, molecular, or atomic scale. (See text box, "Unique Properties Emerge at the Nanoscale.") Accordingly, questions are being raised by representatives of the scientific, advocacy, and regulatory communities about how an EHS research portfolio might be structured when particle size may affect a material's properties, whether it may be necessary to incorporate particle size into regulatory regimes, and how this might be accomplished given the vast spectrum of particle sizes that might affect the characteristics of a particular material.

Some argue that EHS concerns about nanotechnology products can be handled under existing laws and regulations, while others see legal obstacles to adequate EHS regulation. In both of its assessments of the NNI, PCAST concluded that existing regulatory authorities were adequate for the current activities; that appropriate regulatory mechanisms should be used to address instances of harmful human or environmental effects of nanotechnology; and that new regulatory policies related to nanotechnology should be rational, science-based, and consistent across the federal government. Similarly, Sean Murdock, executive director of the NanoBusiness Alliance, asserted that

> The apparatus for effective nanotechnology regulation is largely in place through various statutes and agencies, but it lacks data and resources. To enable these agencies and for the nanotech regulation effort to succeed we must increase the level of funding available to them for nanotech environmental, health and safety research; coordinate efforts between agencies; establish metrics and standards that can be used to characterize nanomaterials; conduct ongoing research; and more.[53]

Others believe that new laws and regulations, or modifications to existing ones, may be required. J. Clarence Davies, senior advisor to the Project on Emerging Nanotechnologies and former EPA Assistant Administrator for Policy, Planning, and Evaluation argued that

> Nanotechnology is difficult to address using existing regulations. There are a number of existing laws — notably the Toxic Substances Control Act; the Occupational Safety and Health Act; the Food, Drug and Cosmetic Act; and the major environmental laws (Clean Air Act, Clean Water Act, and Resource Conservation and Recovery Act) — that provide some legal basis for reviewing and regulating [nanotechnology] materials. However, all of these laws either suffer from major shortcomings of legal authority, or from a gross lack of resources, or both. They provide a very weak basis for identifying and protecting the public from potential risk, especially as nanotechnologies become more complex in structure and function and the applications become more diverse.
>
> A new law may be required to manage potential risks of nanotechnology. The law would require manufacturers to submit a sustainability plan which would show that the product will not present an unacceptable risk.[54]

Davies further asserts that new mechanisms and institutional capabilities — including research programs, tax breaks, acquisition programs, and regulatory incentives — are needed to encourage beneficial applications of nanotechnology.

In developing the regulatory structure, some in the business and financial communities assert that stability and predictability are key characteristics for attracting investment and spurring commercial applications. According to Matthew Nordan, vice president of Lux Research, the

> ambiguity surrounding environmental, health, and safety regulation of nanoparticles is hampering commercialization. Firms do not want to play a game whose rules may change at any time.... That doesn't mean they want more regulations or more onerous regulations. They're just looking for a roadmap on how federal agencies such as the EPA or OSHA [Occupational Safety and Health Administration] plan to approach nanoparticles.[55]

Some tension exists between the goals of promoting the development of nanotechnology, ensuring the global competitive position of the United States, addressing potential EHS implications of nanotechnology, and coping with the unique challenges nanotechnology poses to the current regulatory regime. To prevent health and safety concerns from becoming an impediment to innovation, some suggest that health and safety research and regulation must be done near-concurrently with product development, keeping pace with the speed of innovation. Alternatively, others argue that the potential health, safety, and environmental implications are either unknown or of such significance that EHS research and regulation must precede nanotechnology development and commercialization. "By the time monitoring catches up to commerce the damage will already have been done," asserted Ian Illuminato, health and environment campaigner for Friends of the Earth.[56] AFL-CIO industrial hygienist Bill Kojola warned that

> Even though potential health hazards stemming from exposure have been clearly identified, there are no mandatory workplace measures that require exposures to be assessed, workers to be trained, or control measures to be implemented. [Nanotechnology] should not be rushed to market until these failings are corrected and workers assured of their safety.[57]

The National Research Council assessment of the NNI acknowledged the need for additional reproducible, well-characterized EHS data to inform risk-based guidelines and best practices and warned that until such information is available precautionary measures should be taken to protect the health and safety of workers, the public, and the environment.[58]

In its 2008 assessment of the NNI, PCAST asserted that risk research must not be considered in isolation, but rather in the context of the overall risks and benefits of a particular material or technology. This perspective is shared by many industry advocates who argue that regulatory decisions must balance the potential risks associated with a nanotechnology product against the benefits it delivers and the risk it displaces. Further, they maintain that nanotechnology products should not be held to a higher standard than non-nanotechnology products. PCAST also noted that manufacturers and sellers of nanotechnology products had responsibilities for ensuring workplace and product safety, and asserted that the NNI has a vital role in supporting federal regulatory agencies by providing them with EHS research results.

A description of selected nanotechnology EHS activities of federal regulatory agencies is provided in Appendix A.

International Engagement

International engagement on EHS issues is believed by many to be important to the responsible development and successful commercialization of nanotechnology. NNI officials assert that the United States has played a central role in convening international efforts to address EHS concerns. In its 2008 assessment, PCAST encouraged the NNI to coordinate its efforts with other nations to avoid duplication and to leverage investments, characterizing such work as "non-competitive."[59]

Federal agencies have engaged internationally (e.g., with agencies of other nations, international organizations, standards organizations) across a wide range of nanotechnology-related areas, including standards, nomenclature, and EHS research. Appendix B provides an overview of selected international engagement efforts of NNI agencies related to environmental, health, and safety issues.

Advocates for international engagement assert a variety of potential benefits. For example, transparency and/or harmonization of standards and regulations may contribute to assurance of global supply chains and market confidence in nanotechnology products. Increased globalization of production and markets means that companies and consumers around the world are increasingly part of a common network. Manufacturers of final products generally rely on inputs from multiple suppliers in their global supply chains. The reliability of a final product often depends on the reliability of inputs, such as materials or components. Transparent and common standards and regulations may help to ensure the integrity of supply chains and final products. While this is an issue for a variety of non-nanotechnology products (e.g., the recent discovery of lead-tainted toys and other products imported from China), nanotechnology may present a unique challenge in that at least some nanoscale particles can be incorporated into materials and products in ways that cannot be easily detected or detected at all. Thus, producers and the consumers they serve must rely, in large measure, on standards and regulatory systems to ensure that nanoscale materials are properly produced and represented throughout the supply chain. In the absence of such standards and regulatory systems, producers may not be able to rely on inputs or may incur additional costs for testing and verification; substandard inputs may be incorporated in final products making them underperform or unsafe, and possibly resulting in loss of market confidence and/or potential litigation; or nanotechnology materials may be incorporated without disclosure.

Internationally agreed upon standards could also contribute to greater comparability of research results, improving understanding of EHS-related aspects of nanotechnology, and promoting regulations that help protect human health and the environment. Common standards and nomenclature also may contribute to more effective global collaboration in nanoscale science, engineering, and technology R&D, accelerating the realization of nanotechnology's economic and societal potential.

Global engagement may help to establish a common environment for the development and production of nanotechnology products and to promote access to global markets. In the absence of such an environment, some nations may seek to attract investments in their markets by adopting lower environmental, health, and safety standards and regulations.

Finally, while much remains unknown about the transport and fate of nanoscale materials released into the environment, it is possible that countries and populations other than those where research and production activities take place may be affected. Efforts to promote the adoption of best practices in nanotechnology research, production, use, disposal, and recycling may protect human health and the environment worldwide.

International engagement on EHS research may pose problems, including the time, cost, difficulty, and alleged ineffectiveness of such collaborations. For example, while some advocates assert the need for swift action in advancing EHS research, international engagements often entail slow processes. Also, given the strong U.S. position in nanotechnology, broadly, and in nanotechnology EHS research, specifically, some may argue that other countries have little to contribute, that such efforts tax limited federal EHS financial and human resources, and that such diffusion of resources may slow overall EHS progress. Others might assert that international engagement efforts focused explicitly on nanotechnology are unnecessary given the wide variety of existing mechanisms and pathways for sharing academic research and environmental, health, and safety information across national borders.

Some may oppose international engagement efforts because they lack faith in the goodwill of some participating parties due to the potentially strong national interests at stake (e.g., military applications, economic growth, job creation). In 2003, then- Under Secretary of Commerce for Technology Phillip J. Bond questioned whether global calls for a slowdown in nanotechnology R&D to address environmental, health, and safety concerns are intended to allow other nations to close the nanotechnology leadership gap with the United States:

> I wonder very often if there are really calls for a slow-down so that other governments and countries might catch up.[60]

Others assert that the research required to understand and address EHS implications may be closely linked to applications-related R&D to create nanotechnology materials, products, or processes. In such cases, companies and countries may be reluctant to reveal EHS concerns and efforts, to cooperate in EHS research, or to share results as such actions may reveal competitive strategies, provide information others might use to compete against them (e.g., insights into promising materials or manufacturing processes), or result in unwanted scrutiny by regulators.

Concluding Observations

Advocates and critics agree that potential environmental, health, and safety implications of nanotechnology must be addressed if the full economic and societal benefits of nanotechnology are to be achieved. There is also general agreement that the current body of knowledge of how nanoscale materials might affect humans and the environment is insufficient to assess, address, and manage the potential risks. While there is agreement on the need for more EHS research, there are differing views on the level of funding required, how it should be managed, and related issues.

Congress is currently considering legislation, S. 3274 and H.R. 5940, that would reauthorize and amend the 21st Century Nanotechnology Research and Development Act, the appropriations bills that fund the NNI agencies' nanotechnology EHS research, and two other bills (H.R. 3235, H.R. 4040) with nanotechnology EHS provisions. Congress may use these opportunities to further address nanotechnology EHS implications issues, including: How much should the federal government appropriate for EHS research? How can the federal EHS research investment be better accounted for? How should the research be prioritized? Should the research be more centrally managed? How can EHS research results and best practices be shared more broadly? Can voluntary programs effectively provide needed information about industrial nanotechnology production activities? How can efforts to develop common nomenclature and standards be improved? Are existing laws, regulations, guidelines, and regulatory structures adequate? Is there sufficient coordination among federal regulatory agencies? What types of international engagement on nanotechnology research and regulatory issues could best foster responsible development of nanotechnology?

Nanotechnology EHS-Related Legislation in the 110th Congress

Four bills introduced in the 1 10th Congress contain provisions that seek to address nanotechnology EHS concerns. The following section summarizes selected EHSrelated provisions of these bills.

H.R. 5940 — National Nanotechnology Initiative Amendments Act of 2008

H.R. 5940, the National Nanotechnology Initiative Amendments Act of 2008, was introduced on May 1, 2008. This act would revise the 21st Century Nanotechnology Research and Development Act in a variety of ways, several of which specifically address nanotechnology EHS concerns. The legislation:

- directs the National Nanotechnology Coordination Office to develop and maintain a public database of NNI EHS projects, including the agency funding source and funding history;
- requires the National Nanotechnology Advisory Panel (NNAP) to be established as a "distinct entity" (the NNAP' s functions are currently performed by the President's Council of Advisors on Science and Technology), and requires the establishment of a subpanel to assess whether societal, ethical, legal, environmental, and workforce concerns are adequately addressed by the NNI;
- directs that the National Research Council, as part of its triennial review of the NNI, evaluate the adequacy of the NNI's efforts to address ethical, legal, environmental, human health, and other appropriate societal concerns;

- requires the designation of an associate director of the White House Office of Science and Technology Policy to serve as Coordinator for Societal Dimensions of Nanotechnology with responsibility for developing an annual research plan for federal nanotechnology EHS activities, monitoring and encouraging agency EHS efforts, and for encouraging agencies to engage in public-private partnerships to support EHS research;
- requires certain interdisciplinary research centers supported under the NNI to include EHS research to develop methods for developing environmentally benign nanoscale products and processes, to foster the transfer of research results to industry, and to provide interdisciplinary study programs to educate scientists and engineers in these methods;
- directs NNI agencies to support the activities of standards setting bodies involved in the development of standards for nanotechnology, including authorizing agency reimbursement of travel costs of scientists and engineers participating in these activities; and
- requires activities supported under the NNI' s Education and Societal Dimensions program component area to include environmental, health, and safety education in its informal, pre-college, and undergraduate nanotechnology education efforts.

On June 5, 2008, the House of Representatives passed the bill by a vote of 407-6. The bill has been referred to the Senate Committee on Commerce, Science, and Transportation.

S. 3274 — National Nanotechnology Initiative Amendments Act of 2008

S. 3274, the National Nanotechnology Initiative Amendments Act of 2008, was introduced on July 16, 2008, and referred to the Senate Committee on Commerce, Science, and Transportation. The act would reauthorize and amend the 21st Century Nanotechnology Research and Development Act. The legislation includes a variety of provisions that address nanotechnology EHS issues. The legislation:

- expands the scope of the NNI by identifying as specific missions: (1) support for the development of standard reference materials, instrumentation, measurement science, and computational tools to measure, characterize, and predict the properties of nanoscale materials, and (2) participation in national and international efforts to develop regulatory guidelines, plans, and standards for the safe use of nanotechnology;
- requires, as part of the NNI' s triennial strategic plan, a review of efforts to encourage joint interagency solicitations of grant applications for instrumentation and metrology equipment to detect, measure and characterize nanomaterials;
- directs NNI agencies to support the activities of standards setting bodies involved in the development of standards for nanotechnology, including authorizing agency reimbursement of travel costs of scientists and engineers participating in these activities;

- directs the National Nanotechnology Coordination Office to develop and maintain a public database of NNI EHS projects, including a project description, agency funding source, and funding history;
- requires the National Nanotechnology Advisory Panel (NNAP) to be established as a "distinct entity" (the NNAP' s functions are currently performed by the President's Council of Advisors on Science and Technology), and requires the establishment of a subpanel on societal, ethical, legal, environmental, and workforce concerns;

In: Nanotechnology in the USA: Developments, Policies and Issues ISBN 978-1-60692-800-4
Editor: Carl H. Jennings

Chapter 4

NANOTECHNOLOGY: BETTER GUIDANCE IS NEEDED TO ENSURE ACCURATE REPORTING OF FEDERAL RESEARCH FOCUSED ON ENVIRONMENTAL, HEALTH, AND SAFETY RISKS*

United States Government Accountability Office

WHY UNITED STATES GOVERNMENT ACCOUNTABILITY OFFICE DID THIS STUDY

The National Nanotechnology Initiative (NNI), administered by the Office of Science and Technology Policy (OSTP), is a multiagency effort intended to coordinate the nanotechnologyrelated activities of 25 federal agencies that fund nanoscale research or have a stake in the results. Nanotechnology is the ability to control matter at the scale of a nanometer—one billionth of a meter. A key research area funded by some federal agencies relates to potential environmental, health, and safety (EHS) risks that may result from exposure to nanoscale materials. Because of concerns about federal efforts to fund and prioritize EHS research, GAO was asked to determine (1) the extent to which selected agencies conducted such research in fiscal year 2006; (2) the reasonableness of the agencies' and the NNI's processes to identify and prioritize such federal research; and (3) the effectiveness of the agencies' and the NNI's process to coordinate this research. GAO reviewed quantitative and qualitative data from five federal agencies that provided 96 percent of fiscal year 2006 funding for EHS research.

* This is an edited, excerpted and augmented edition of a United States Government Accountability Office Report GAO-08-402, Dated March 2008.

What United States Government Accountability Office Recommends

GAO is recommending that OSTP provide better guidance to agencies regarding how to report research that is primarily focused on EHS risks. In commenting on a draft of this report, OSTP generally agreed with the findings and will review the manner in which agencies respond to current guidance.

What United States Government Accountability Office Found

The NNI reported that in fiscal year 2006, federal agencies devoted \$37.7 million—or 3 percent of the \$1.3 billion total nanotechnology research funding—to research that was primarily focused on the EHS risks of nanotechnology. However, about 20 percent of this total cannot actually be attributed to this purpose; GAO found that 22 of the 119 projects identified as EHS-related by five federal agencies in fiscal year 2006 were not focused on determining the extent to which nanotechnology poses an EHS risk. Instead, the focus of many of these projects was to explore how nanotechnology could be used to remediate environmental damage or to detect a variety of hazards. GAO determined that this mischaracterization is rooted in the current reporting structure which does not allow these types of projects to be easily categorized and the lack of guidance for agencies on how to apportion funding across multiple topics. In addition to the EHS funding totals reported by the NNI, federal agencies conduct other research that is not captured in the totals. This research was not captured by the NNI because either the research was funded by an agency not generally considered to be a research agency or because the primary purpose of the research was not to study EHS risks.

Federal agencies and the NNI are currently in the process of identifying and prioritizing EHS risk research needs; the process they are using appears reasonable overall. For example, identification and prioritization of EHS research needs is being done by the agencies and the NNI. The NNI also is engaged in an iterative prioritization effort through its Nanotechnology Environmental and Health Implications (NEHI) working group. NEHI has identified five specific research priorities for five general research categories, but it has not yet completed the final steps of this process, which will identify EHS research gaps, determine specific research needed to fill those gaps, and outline a long-term, overarching EHS research strategy. GAO found that the focus of most EHS research projects underway in fiscal year 2006 was generally consistent with agency priorities and NEHI research categories and that the projects focused on the priority needs within each category to varying degrees. The anticipated EHS research strategy is expected to provide a framework to help ensure that the highest priority needs are met.

Agency and NNI processes to coordinate activities related to potential EHS risks of nanotechnology have been generally effective. The NEHI working group has convened frequent meetings that have helped agencies identify opportunities to collaborate on EHS risk issues, such as joint sponsorship of research and workshops to advance knowledge and facilitate information- sharing among the agencies. In addition, NEHI has incorporated several practices that are key to enhancing and sustaining interagency collaboration, such as leveraging resources. Finally, agency officials GAO spoke with expressed satisfaction with the coordination and collaboration on EHS risk research that has occurred through NEHI. They cited several factors they believe contribute to the group's effectiveness, including the stability of the working group membership and the expertise and dedication of its members.

ABBREVIATIONS

CPSC	U.S. Consumer Product Safety Commission
EHS	environmental, health, and safety
EPA	Environmental Protection Agency
FDA	Food and Drug Administration
HHS	Department of Health and Human Services
NEHI	Nanotechnology Environmental and Health Implications (working group)
NIH	National Institutes of Health
NIOSH	National Institute for Occupational Safety and Health
NIST	National Institute of Standards and Technology
NNCO	National Nanotechnology Coordinating Office
NNI	National Nanotechnology Initiative
NSET	Nanoscale Science, Engineering, and Technology (subcommittee)
NSF	National Science Foundation
NSTC	National Science and Technology Council
OMB	Office of Management and Budget
OSHA	Occupational Safety and Health Administration
OSTP	Office of Science and Technology Policy
PCA	program component area

March 31, 2008
The Honorable Daniel K. Inouye, Chairman
The Honorable Mark L. Pryor
The Honorable Gordon H. Smith
Committee on Commerce, Science, and Transportation United States Senate
The Honorable Ron Wyden
The Honorable Richard Burr
Congressional Nanotechnology Caucus United States Senate

Nanotechnology encompasses a wide range of innovations based on the understanding and control of matter at the scale of nanometers—the equivalent of one-billionth of a meter. For illustration, a sheet of paper is about 100,000 nanometers thick, a human hair is about 80,000 nanometers wide, and 2 gold atoms lying side by side are about 1 nanometer long. At the nanoscale level, some materials may exhibit electrical, magnetic, biological, and other properties that differ significantly from properties the same materials exhibit at a larger scale. For example, opaque materials, such as copper, become transparent at the nanoscale and inert materials, such as platinum and gold, become chemical catalysts. Exploiting the differences in the size and properties of materials at the nanoscale level has led to a range of commercial uses and holds the promise for innovations in virtually every industry from aerospace and energy to health care and agriculture. In 2006, an estimated $50 billion in products worldwide incorporated nanotechnology and this figure has been projected to grow to $2.6 trillion by 2014, according to an industry analyst. The Woodrow Wilson International Center for Scholars' Project on Emerging Nanotechnologies has identified over 500 consumer products that already are available to consumers that may contain nanoscale materials. For example, nanoscale materials have been embedded into clothing fabric to repel stains. Some nanoscale materials in development could greatly improve pharmaceuticals because the materials' size, structure, and behavior can be used, for example, to treat diseases by delivering drugs directly to affected cells, such as tumor cells. Food companies also are experimenting with nanoscale materials that can be incorporated into food packaging to detect spoilage or pathogens, and cosmetics companies have developed products with nanoscale materials that reportedly enable sunscreens to perform better.

While the use of nanoscale materials holds much promise for the future, the small size and unique properties of nanomaterials raise questions about potential environmental, health, and safety (EHS) effects—referred to as EHS risks—that might result from exposures during the manufacture, use, and disposal or recycle of nanoscale materials. For example, recent research suggests that nanomaterials are small enough to get inside cells and some may cross the blood-brain barrier to directly enter the central nervous system. Because nanotechnology is a relatively new science, basic information about the properties of many nanoscale materials is not fully known. Scientists are working to fill significant gaps in current knowledge about nanoscale materials so they can answer questions about potential EHS risks and help ensure the safe commercial development of these materials.

In 2001, the National Nanotechnology Initiative (NNI) was established as a federal, multiagency effort intended to accelerate the discovery, development, and deployment of nanoscale science, engineering, and technology to achieve economic benefits, enhance the quality of life, and promote national security.[1] The NNI is a mechanism to coordinate the nanotechnology-related activities of the 25 currently participating federal agencies that fund nanoscale research or have a stake in the outcome of this research, such as those agencies that may regulate products containing nanomaterials.[2] While the NNI is designed to facilitate intergovernmental cooperation and identify overarching goals and priorities for nanotechnology research, it is not a research program. It also has no funding or authority to dictate the nanotechnology research agenda for participating agencies or to ensure that adequate resources are available to achieve specific goals. Instead, participating agencies develop and fund their own nanotechnology research agendas. In fiscal year 2006, 13 of the 25 participating agencies in the NNI allocated a total of about $1.3 billion from their appropriated budgets to nanotechnology research and development activities.

Management of the NNI falls under the purview of the National Science and Technology Council (NSTC), a Cabinet-level body within the Office of the President that coordinates science and technology policy across the federal government. The NSTC's Committee on Technology has established a Nanoscale Science, Engineering, and Technology (NSET) subcommittee to help coordinate, plan, and implement the NNI's activities across participating agencies. In 2003, the NSET subcommittee further established a Nanotechnology Environmental and Health Implications (NEHI) working group.[3] The purpose of the NEHI working group, composed of representatives from 16 research and regulatory agencies, is to, among other things, coordinate agency efforts related to EHS risks of nanotechnology. As is the case with the NNI, the NEHI working group has no authority to mandate research priorities or to ensure that agencies adequately fund particular research.

In December 2003, Congress enacted the 21st Century Nanotechnology Research and Development Act.[4] The act establishes a National Nanotechnology Program to coordinate federal nanotechnology research and development. Among other things, the act directs the NSTC to establish goals and priorities for the program and to set up program component areas that reflect those goals and priorities. To implement these requirements, the NSTC has established a process to categorize research projects and activities undertaken by the various federal agencies into seven areas. Of these seven, six are focused on the discovery, development, and deployment of nanotechnology, while the seventh relates to the societal dimensions of nanotechnology that include issues such as the EHS risks of nanotechnology. Agencies also report their research funding for each area to the Office of Management and Budget (OMB) as part of the annual federal budget process. NNI's annual Supplement to the President's Budget, prepared by the NSTC, includes EHS research figures from the agencies and a general description of the research conducted by the agencies in each of the seven areas. For reporting purposes, the NSET subcommittee has defined EHS research as "efforts whose primary purpose is to understand and address potential risks to health and to the environment posed by this technology." Eight of the 13 agencies that dedicated a portion of their research budgets for nanotechnology research in fiscal year 2006 reported having devoted some resources to research that had a primary focus on potential EHS risks. However, the Woodrow Wilson International Center for Scholars has questioned the accuracy of reporting research related to EHS risks. Furthermore, some groups, including industry, environmental advocacy, and nonprofit research institutes, have raised concerns about the pace of NEHI's prioritization activities as well as the process it is using to identify research priorities and coordinate federal research.

In this context you asked us to report on (1) the extent to which selected research and regulatory agencies conducted research in fiscal year 2006 that primarily was focused on the potential EHS risks of nanotechnology; (2) the reasonableness of the processes that agencies and the NNI use to identify and prioritize federal research on the potential EHS risks of nanotechnology; and (3) the effectiveness of the processes that agencies and the NNI use to coordinate their research.

To determine the extent to which selected research and regulatory agencies conducted research that is primarily focused on studying the EHS risks of nanotechnology, we gathered data on the funding that NNI's participating agencies have used for EHS risk research. We focused our review on the Environmental Protection Agency (EPA), the National Institutes of Health (NIH), the National Institute for Occupational Safety and Health (NIOSH), the National Institute of Standards and Technology (NIST), and the National Science Foundation (NSF), because these five agencies accounted for 96 percent of the EHS research funding reported in fiscal year 2006. Of these agencies, NIH, NIOSH, NIST, and NSF are research agencies that have specific budgets to support research, including nanotechnology-related research. EPA on the other hand is a regulatory agency that also conducts research and therefore has a research budget. In addition to the agencies mentioned above, we also included in our review three regulatory agencies that do not have research budgets—the U.S. Consumer Product Safety Commission (CPSC), the Food and Drug Administration (FDA), and the Occupational Safety and Health Administration (OSHA) to determine whether these three agencies conducted any research on their own relative to EHS risks of nanotechnology. We assessed the reliability of the agencies' data and determined it was sufficient for the purposes of this analysis. To assess whether or not the primary purpose of the research conducted by these agencies addressed the EHS risks of nanotechnology, we reviewed qualitative data on all projects funded by EPA, NIH, NIOSH, NIST, and NSF in fiscal year 2006. To minimize bias and to ensure the consistency of our evaluation, the team independently conducted project reviews by using publicly available and agency documentation, such as project abstracts or grant applications, to make our determinations. For categorization of projects that appeared questionable to us, we discussed the categorization with agency officials and modified our determination as appropriate given the additional support provided by the agency.

To determine the reasonableness of the process that the agencies and the NNI used to prioritize and coordinate federal research on studying EHS risks, we collected and reviewed documentation on research priorities, and the process used to establish these priorities, at each of the eight agencies included in our review and compared these priorities with funded research within the agency. To review the process being used by the NNI to identify and establish government-wide priorities, we reviewed NNI documents and interviewed agency officials and external stakeholder groups, including officials from groups that represent environmental and industry concerns. We compared the NNI's identified priorities with those identified by the agencies to determine whether they were consistent. We also compared the NNI's identified priorities with agency project-level data on EHS research underway in fiscal year 2006 to determine whether the projects were reflective of NNI's identified priorities. We did not determine whether the NNI's identified priorities represented a scientific consensus on the most appropriate ones. We interviewed agency officials at each of the eight agencies about the extent to which their agency's research priorities were met, either through the agency's own research or research conducted by other agencies. With regard to coordination, we discussed with agency and NNI officials how agencies coordinate research and NNI's role to facilitate that coordination, and we obtained documentation on these collaborative efforts. Furthermore, we compared the NNI's efforts to facilitate interagency collaboration with established practices that have been found to enhance and sustain collaboration among federal agencies. In addition, we interviewed stakeholders, including environmental and industry groups, to obtain views on agency coordination efforts. We conducted this performance audit from June 2007 to February 2008 in accordance with generally accepted government auditing standards. Those standards require that we plan and perform the audit to obtain sufficient, appropriate evidence to provide a reasonable basis for our findings and conclusions based on our audit objectives. We believe that the evidence obtained provides a reasonable basis for our findings and conclusions based on our audit objectives.

Results in Brief

Of the $1.3 billion that federal agencies allocated to nanotechnology research in fiscal year 2006, the NNI reported that about $37 million was devoted to research that primarily focused on studying the EHS risks of nanotechnology. However, based on our analysis, about one-fifth of this amount cannot actually be attributed to this purpose. Specifically, our analysis found that 22 of the 119 projects identified as EHS projects by EPA, NIH, NIOSH, NIST, and NSF in fiscal year 2006 were not primarily related to understanding the EHS risks of nanotechnology. These 22 projects, funded by NSF and NIOSH, accounted for about $7 million of the $37 million that the NNI reported as being primarily focused on EHS risks. Instead of determining the extent to which nanotechnology poses an EHS risk, the primary purpose of many of these projects was to explore how nanotechnology could be applied to remediate environmental damage or could be used to detect a variety of hazards, such as chemical or biological. The miscategorization of these 22 projects results largely from a reporting structure for nanotechnology research that does not easily allow agencies to recognize projects that use nanotechnology to mitigate environmental damage or enhance detection of environmental contaminants and from the lack of guidance available to the agencies on how to apportion funding across multiple topics, when appropriate. As a result, agency officials said they characterized these projects as being EHS-focused for lack of a more closely related category to place them in. We also determined that some federal agencies conduct research that is not reported as part of EHS research funding and is therefore not captured in the EHS totals provided by the NNI. For example, NIH has research underway to develop drug delivery mechanisms that use nanotechnology. This research also will provide relevant health and safety information on how nanomaterials interact with the body at the cellular level, but the agency's funding for this type of research is not included in the NNI's totals for EHS research because studying EHS risks was not its primary purpose. We are recommending that the Office of Science and Technology Policy (OSTP), in consultation with the NNI and OMB, provide better guidance to agencies regarding how to report research that is primarily focused on understanding or addressing the EHS risks of nanotechnology. In commenting on this chapter, OSTP generally concurred with the chapter's findings and agreed to review the manner in which agencies respond to the current guidance at future NSET meetings.

The agencies and the NNI are currently in the process of identifying and prioritizing EHS risk research needs; the process they are using appears reasonable overall. Identification and prioritization of research needs related to EHS risks takes place within individual agencies as well as within the NNI. Agencies' priorities are linked to their missions and are generally set by intra-agency teams dedicated to nanotechnology issues. Most of the eight agencies we reviewed have established internal task forces to identify and prioritize nanotechnology research needs and to communicate these priorities to the larger research community. In addition to these agency efforts, the NSET subcommittee is currently engaged in an iterative prioritization effort through its NEHI working group. This effort began with a September 2006 report in which NEHI identified five general categories of research necessary to evaluate EHS risks and a list of 75 specific research needs, which were not prioritized at that time. Subsequently, in an August 2007 report, NEHI distilled the list of 75 specific research needs into a set of five prioritized needs under each of the five general research categories. Agency officials told us that NEHI's report generally reflects their agencies' key research priorities. NEHI has not yet completed the final steps of this process and plans to issue a report in early 2008 that will identify EHS research gaps; determine specific research needed to fill those gaps; and outline a long-term, overarching strategy to guide agency research funding decisions. Furthermore, our analysis of the 97 research projects that were underway in fiscal year 2006 that were primarily related to studying EHS risks found that the focus of these projects was generally consistent with agency priorities and NEHI's five general research categories and that the projects focused on the priority needs within each category to varying degrees. The anticipated 2008 NEHI report is expected to provide a framework to help agencies better target the highest priority research needs in the future. Also, some environmental and industry groups have advocated for a more top-down and directed approach for setting and funding federal nanotechnology research priorities. However, such a structure and approach is generally inconsistent with historical approaches used to set federal research priorities and may be difficult to implement given how federal research is currently funded.

Agency and NNI processes to coordinate activities related to the potential EHS risks of nanotechnology have been generally effective. The NEHI working group has convened frequent meetings, augmented by informal discussions among agencies, that have helped agencies identify opportunities to collaborate on EHS risk issues. These interagency collaborations have taken many forms including joint sponsorship of EHSrelated research and workshops and detailing staff to work at other NEHI participating agencies. These types of exchanges, according to most agency officials we spoke with, have helped advance knowledge and facilitated information-sharing among the agencies. In addition, NEHI has incorporated several practices that we have previously identified as key to enhancing and sustaining interagency collaborative efforts, such as defining a common outcome and leveraging resources, but has not completed an overarching strategy to help better align agencies' EHS research efforts. Finally, all agency officials we spoke with expressed satisfaction with both the coordination and the collaboration on EHS risk research that has occurred through NEHI. These officials cited several factors that they believe have contributed to the working group's effectiveness, including the expertise, dedication, and low turnover rate of its members. Furthermore, according to these officials, this stability, combined with common research needs and general excitement about the new science, has resulted in a collegial, productive working environment.

BACKGROUND

Nanotechnology is generally defined as the ability to understand and control matter at the nanoscale (between 1 and 100 nanometers), in order to create materials, devices, and systems with fundamentally new properties and functions specific to that scale. For example, opaque materials, such as copper, become transparent at the nanoscale and inert materials, such as platinum and gold, become chemical catalysts. With the capacity to control and manipulate matter at this scale, nanotechnology promises advances in areas such as new drug delivery systems, more resilient materials and fabrics, stronger materials at a fraction of the weight, more efficient energy conversion, and dramatically faster computer chips.

To guide federal development of this technology, the National Nanotechnology Initiative (NNI) was established in fiscal year 2001 to support long-term research and development aimed at accelerating the discovery, development, and deployment of nanoscale science, engineering, and technology. The NNI is a multiagency program involving nanotechnology-related activities of the 25 federal agencies currently participating, including the National Science Foundation (NSF), the Department of Defense, the Department of Energy, the

National Institutes of Health (NIH), and the National Institute of Standards and Technology (NIST). See table 1 for a complete listing of federal agencies participating in the NNI as of December 2007.

Table 1. Federal Agencies Participating in the National Nanotechnology Initiative, as of December 2007

Federal Agencies with Budgets Dedicated to Nanotechnology Research and Development	Other Participating Agencies
Cooperative State Research, Education, and Extension Service Department of Defense Department of Energy Department of Homeland Security Department of Justice Department of Transportation Environmental Protection Agency National Aeronautics and Space Administration National Institute of Standards and Technology National Institute for Occupational Safety and Health National Institutes of Health National Science Foundation U.S. Forest Service	Bureau of Industry and Security Department of Education Department of Labor Department of State Department of the Treasury Food and Drug Administration International Trade Commission Intelligence Advanced Research Projects Activity Nuclear Regulatory Commission U.S. Consumer Product Safety Commission U.S. Geological Survey U.S. Patent and Trademark Office

Source: NNI.

Federal support for nanotechnology research totaled about $1.3 billion in fiscal year 2006. Cumulatively through fiscal year 2006, federal agencies have devoted over $5 billion to nanotechnology research since the NNI's inception. While not all of the NNI's participating agencies conduct or sponsor research, in fiscal year 2006, 13 agencies had budgets dedicated to nanotechnology research and development. Eight of these 13 agencies devoted some of their research resources to studying the environmental, health, and safety (EHS) risks of nanotechnology. Of these eight agencies, five—EPA, NIH, NIOSH, NIST, and NSF—accounted for almost 96 percent of the research focused on EHS risks in fiscal year 2006. NSF alone accounted for about 56 percent of all federal EHS risk research in fiscal year 2006. See figure 1 for a break out of research funds used by agency.

A number of research and regulatory agencies support research to advance knowledge and information about the potential EHS risks of nanotechnology:

- The National Institute for Occupational Health and Safety (NIOSH) is a research agency within the Department of Health and Human Services (HHS) that concentrates its research on topics related to human health. NIOSH's research results in recommendations for preventing work-related injuries, illnesses, and death. It therefore focuses on studies that will improve scientists' ability to identify potential adverse occupational health effects of nanomaterials.

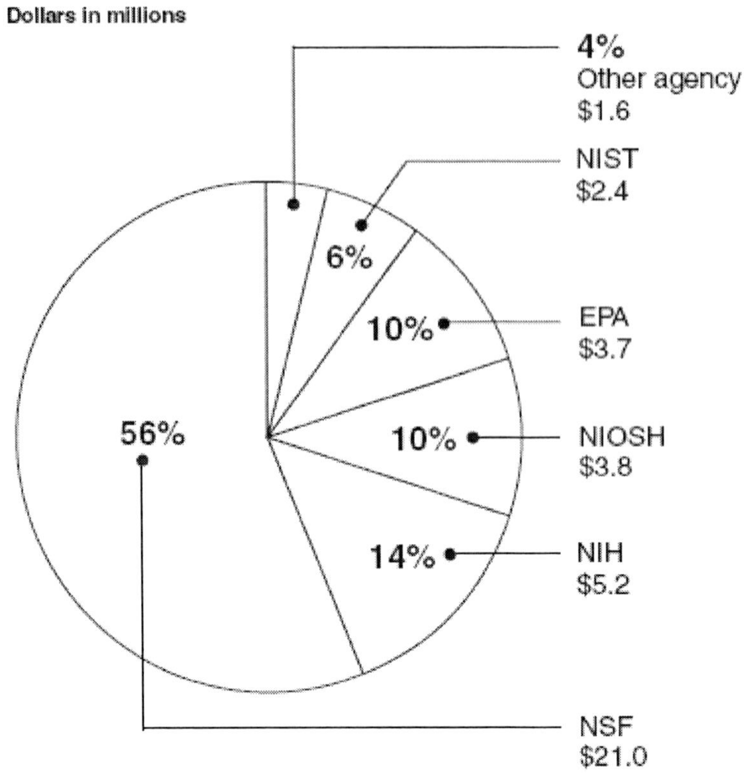

Source: GAO analysis of NNI budget data

Figure 1. Nanotechnology EHS Research by Agency, as Reported by the National Nanotechnology Initiative, Fiscal Year 2006.

- At NIH, another HHS research agency that concentrates on human health, nanotechnology research is generally focused on the development of medical applications and the protection of public health, including research to examine the interaction of nanomaterials with biological systems.
- Consistent with its mission to advance measurement science, standards, and technology to enhance economic security and improve our quality of life, the National Institute of Standards and Technology (NIST), an agency in the Department of Commerce, develops the measurement techniques required to better characterize potential impacts of nanotechnology.
- The National Science Foundation (NSF) has the broadest research portfolio relative to nanotechnology and supports research to help meet its mission to promote the progress of science and engineering. With regard to EHS risks, NSF sponsors research to develop new methods to characterize nanoparticles and investigate the environmental implications and toxicity of nanomaterials. In addition, NSF sponsors a network of research centers that focus on a range of EHS issues including occupational safety during nanomanufacturing and the interaction of nanomaterials and cells.
- In addition to these research agencies, a number of regulatory agencies also have an interest in developing information about the potential EHS risks of nanotechnology:

- The Environmental Protection Agency (EPA), which is both a research and regulatory agency, is tasked with protecting human health and the environment. As a result, EPA determined that it needed to develop a better understanding of the potential human health and environmental risks from exposure to nanoscale materials and is therefore focusing its research efforts in this area, among others.
- The Food and Drug Administration (FDA), another HHS agency, is generally responsible for overseeing the safety and effectiveness of drugs and devices for humans and animals, and of biological products for humans. The agency also is generally responsible for overseeing the safety of color additives, cosmetics, and foods, including food additives and dietary supplements. As a result, FDA is interested in understanding the potential risks posed by nanomaterials used in products under its jurisdiction.
- The Occupation Safety and Health Administration (OSHA) is a Department of Labor agency whose mission is, in part, to ensure the safety and health of workers by setting and enforcing standards and encouraging continual improvement in workplace safety and health. OSHA is interested in information that would aid in the application of existing health standards—including hazard communication, respiratory protection programs, and laboratory standards—to nanotechnology operations and help determine the need for new standards or guidance products.
- • The mission of the U.S. Consumer Product Safety Commission (CPSC) is to protect the public from unreasonable risks of serious injury or death from more than 15,000 types of consumer products, including some that may be manufactured with nanomaterials.
- The NNI is managed within the framework of the National Science and Technology Council's (NSTC) Committee on Technology. The NSTC is an organization through which the President coordinates science and technology policies across the federal government. The NSTC is managed by the Director of the Office of Science and Technology Policy (OSTP), who also serves as the Science Advisor to the President. The NSTC's Committee on Technology established the Nanoscale Science, Engineering, and Technology (NSET) subcommittee to coordinate communication between the federal government's multiagency nanoscale research and development programs. The NSET subcommittee is composed of representatives from any agencies that choose to participate in the NNI (as of January 2008, 25 agencies are involved) and serves as the primary interagency coordination mechanism for nanotechnology-related research. Supporting the NSET subcommittee, the National Nanotechnology Coordinating Office (NNCO) provides day-to-day technical guidance and administrative assistance to prepare multiagency planning, budget, and assessment documents. In addition, the NSET subcommittee has established a number of working groups to help better focus interagency attention and activity on specific issues, such as the Nanotechnology Environmental and Health Implications (NEHI) working group. This group was designed to provide for exchange of information among participating agencies; facilitate the identification, prioritization, and implementation of research; and promote communication to other federal and nonfederal entities. The NEHI working group also coordinates U.S. participation in international activities, including the programs of the Organisation for Economic Co-

operation and Development. Currently, NEHI membership consists of 16 research and regulatory agencies. See figure 2 for the NNI's structure.

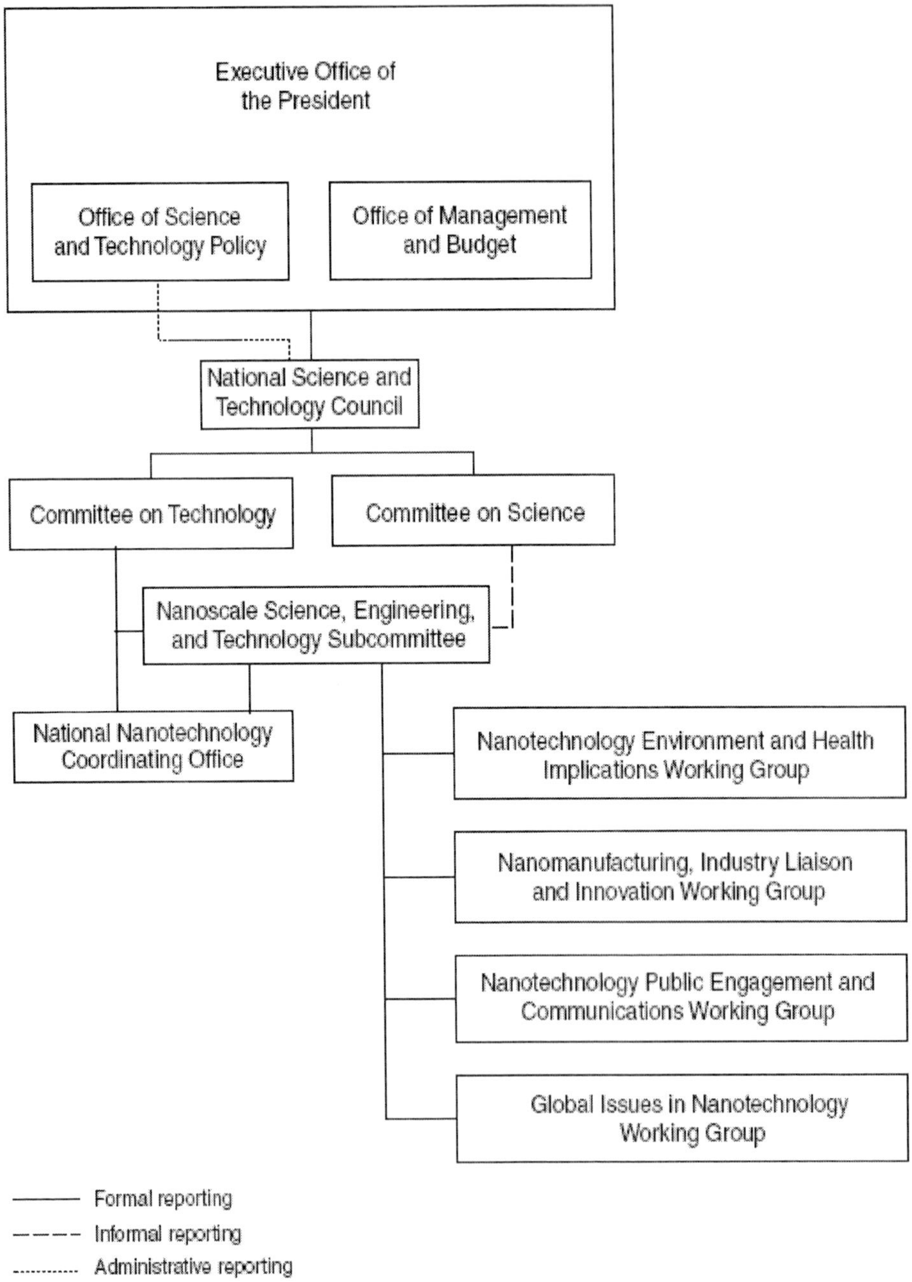

Source:NNL.

Figure 2. National Nanotechnology Initiative Structure.

Under the NNI, each agency funds research and development projects that support its own mission as well as the NNI's goals. While agencies share information on their nanotechnology-related research goals with the NSET subcommittee and NEHI working group, each agency retains control over its decisions on the specific projects to fund. While the NNI was designed to facilitate intergovernmental cooperation and identify goals and priorities for nanotechnology research, it is not a research program. It has no funding or authority to dictate the nanotechnology research agenda for participating agencies.

The NNI used its fiscal year 2000 strategic plan and its subsequent updates to delineate a strategy to support long-term nanoscale research and development, among other things. A key component of the 2000 plan was the identification of nine specific research and development areas— known as "grand challenges"—that highlighted federal research on applications of nanotechnology with the potential to realize significant economic, governmental, and societal benefits.[5] Examples of potential breakthroughs cited in this strategic plan included developing materials that are 10 times stronger, but significantly lighter, than steel to make vehicles lighter and more fuel efficient; improving the speed and efficiency of computer transistors and memory chips by factors of millions; and developing methods to detect cancerous tumors that are only a few cells in size using nanoengineered contrast agents.

In 2004, the NNI updated its strategic plan and described its goals as well as the investment strategy by which those goals were to be achieved.[6] Consistent with the 21st Century Nanotechnology Research and Development Act, the NNI established major subject categories of research and development investment, called program component areas (PCA), that cut across the interests and needs of the participating agencies.[7] These seven areas replaced the nine grand challenges and other nanotechnology investment areas that the agencies had previously used to categorize their nanotechnology research. Six of the seven areas are focused on the discovery, development, and deployment of nanotechnology. The seventh, societal dimensions, consists of two subareas—research on environmental, health, and safety; and education and research on ethical, legal, and other societal aspects of nanotechnology. The EHS portion of the societal dimensions PCA accounted for over $37 million in fiscal year 2006. See figure 3 for a break out of research funds used, by PCA.

PCAs are intended to provide a means by which the NSET subcommittee, OSTP, the Office of Management and Budget (OMB), Congress, and others may be informed of the relative federal investment in these key areas. PCAs also provide a structure by which the agencies that fund research and development can better direct and coordinate their activities. In response to increased concerns about the potential EHS risks of nanotechnology, in fiscal year 2005, the NSET subcommittee and the agencies agreed to separately report their research funding for each of the two components of the societal dimensions PCA. The December 2007 update of the NNI's strategic plan reaffirmed the program's goals, identified steps to accomplish those goals, and formally divided the societal dimensions PCA into two separate PCAs—"environment, health, and safety" and "education and societal dimensions."

Beginning with the development of the fiscal year 2005 federal budget, agencies have worked with OMB to identify funding for nanoscale research that would be reflected in the NNI's annual Supplement to the President's Budget. Specifically, OMB issued guidance that consisted of a definition of nanoscale research and a notice that OMB would work with agencies to identify data for each of the PCAs. OMB analysts reviewed aggregated, rather than project-level, data on research funding for each PCA to help ensure consistent reporting across the agencies.

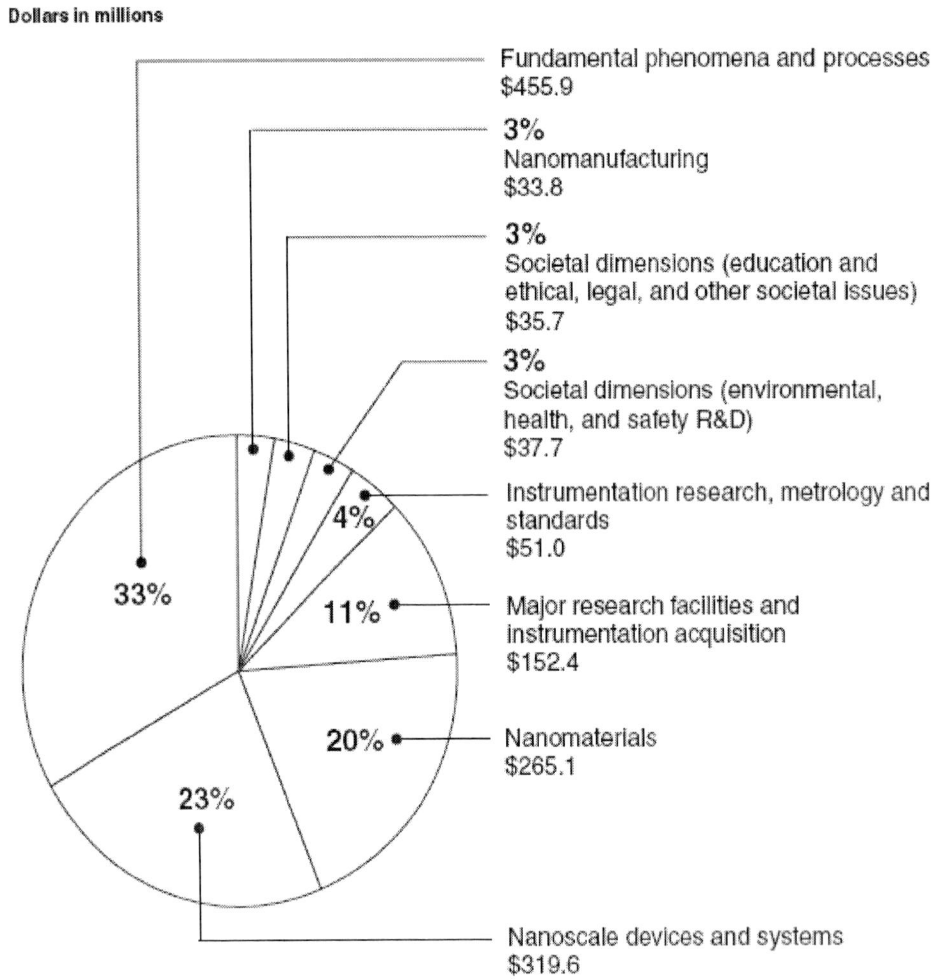

Source: GAO analysis of NNI budget data.

Figure 3. Nanotechnology Research Funding by Program Component Area, Fiscal Year 2006

Agencies also relied on definitions of the specific PCAs developed by the NSET subcommittee to determine the appropriate area in which to report research funding. Neither NSET nor OMB provided guidance on whether or how to apportion funding for a single research project to more than one PCA, if appropriate. However, representatives from both NSET and OMB stressed that the agencies were not to report each research dollar more than once.

Almost 20 Percent of EHS Research Projects were not Primarily Focused on Studying The EHS Risks of Nanotechnology

Although the NNI reported that federal agencies in fiscal year 2006 devoted $37.7 million—or about 3 percent of the total of all nanotechnology research funding—to research that primarily focused on studying the EHS risks of nanotechnology, we found that about 18 percent of the EHS research reported by the NNI cannot actually be attributed to this purpose. This was largely due to a reporting structure that did not lend itself to categorizing particular types of projects and limited guidance provided to the agencies by the NNI on how to consistently report EHS research. In addition to research reported as being primarily focused on the EHS risks of nanotechnology, some agencies conduct research that is not reflected in the EHS totals provided by the NNI either because they are not considered federal research agencies or because the primary purpose

EHS Research Constituted about 3 Percent of Federal Nanotechnology Research Funding in Fiscal Year 2006

Overall, 3 percent—or $37.7 million—of the approximately $1.3 billion dedicated for nanotechnology research funding in fiscal year 2006 was reported as being devoted to studying the EHS risks of nanotechnology. Our review of data on agency funding for 119 projects that were underway in fiscal year 2006 largely confirmed the figures reported by the NNI. Specifically, all but one of the five individual agencies reported the same or greater funding to us than what the NNI reported for fiscal year 2006. EPA reported slightly less to us than it did to the NNI. Largely these discrepancies resulted from timing differences in the date the NNI needed the data and the date agency officials finalized their review of fiscal year spending. For example, NIOSH reported $470,000 more to us because it had not included funding for a few projects in its report to the NNI, according to agency officials. Other differences resulted from rounding.

As would be expected, our review of the descriptive information on EHS projects found that those agencies with missions directly related to protecting the environment or human health and safety devoted a greater percentage of their nanotechnology research budgets to studying EHS risks. For example, in fiscal year 2006, NIOSH reported devoting 100 percent of its fiscal year 2006 nanotechnology research funds to support 23 projects to study EHS risks. These projects focused primarily on worker safety and exposure, such as gathering data on workplace exposure to nanomaterials and evaluating the extent to which particle size affects the toxicity of inhaled nanomaterials. Similarly, EPA reported devoting 82 percent of its nanotechnology research budget to study EHS risks. This research included human health-focused projects to examine the toxicity of manufactured nanomaterials at the molecular and cellular level, as well as environmentally focused projects to evaluate how nanomaterials disperse and change under different environmental conditions and the extent to which nanomaterials accumulate in the bodies of various animal species.

In contrast, we found that agencies with broader missions devoted a smaller portion of their nanotechnology research funds to study EHS issues. For example, NIST, an agency oriented toward measurement science and standards, dedicated 3 percent of its nanotechnology research budget to EHS risks in fiscal year 2006. The majority of its research funding focused on such PCAs as fundamental phenomena and processes; nanoscale devices and systems; and instrumentation research, metrology, and standards. Similarly, NSF dedicated 6 percent of its fiscal year 2006 nanotechnology research funds on research related to EHS risks as compared with 41 percent focused on fundamental phenomena and processes.

In fiscal year 2008, funding for both EHS-related research and nanoscale research in general is projected to grow. Overall nanotechnology research is projected to increase in fiscal year 2008 to about $1.4 billion, or an increase of 20 percent over fiscal year 2005 figures. Funding for EHSrelated research is expected to increase to approximately $59 million, an increase of 68 percent over fiscal year 2005 levels. As a result, EHS research would grow to about 4 percent of projected nanotechnology research in fiscal year 2008.

CURRENT REPORTING STRUCTURE AND LIMITED GUIDANCE CONTRIBUTE TO INACCURATE REPORTING OF EHS RISK RESEARCH

About 18 percent of the total research dollars reported by the agencies as being primarily focused on the study of nanotechnology-related EHS risks in fiscal year 2006 cannot actually be attributed to this purpose. Specifically, our analysis found that 22 of the 119 projects funded by five federal agencies were not primarily related to studying EHS risks. These 22 projects accounted for about $7 million of the total that the NNI reported as supporting research primarily focused on EHS risks. Almost all of these projects—20 out of 22—were funded by NSF, with the two additional projects funded by NIOSH. See table 2 for our analysis of the nanotechnology research projects reported as being primarily focused on EHS risks.

Table 2. GAO Analysis of the Number and Dollar Value of Nanotechnology Research Projects Reported by Selected Agencies as Being Primarily Focused on Environmental, Health, and Safety Risks, Fiscal Year 2006

(Dollars in millions)

	Projects reported by agencies as being primarily		Projects determined by GAO to be primarily		Projects determined by GAO not to be primarily focused on	
Agency	Number	Dollar Value[a]	Number	Dollar Value	Number	Dollar value
EPA	10	$3.6	10	$3.6	0	$0
NIH	18	$5.6	18	$5.6	0	$0
NIOSH	23	$4.3	21	$4.2	2	$0.1
NIST	2	$2.4	2	$2.4	0	$0
NSF	66	$21.1	46	$14.7	20	$6.4
Total	119	$37	97	$30.5	22	$6.5

Source: GAO analysis of agency obligations data.

a. Figures differ slightly from those reported by the NNI in the Supplement to the President's FY2008 Budget due to rounding error or modifications made to the project-level data after they were reported by agencies to the NNI.

We found that the primary purpose of many of these 22 projects was to explore ways to use nanotechnology to remediate environmental damage or to identify environmental, chemical, or biological hazards. For example, a number of NSF projects explored the use of nanotechnology to improve water or gaseous filtration systems. In other cases, NSF-funded research was targeted toward developing nanotechnology-based applications to remediate soil or water contamination. In addition, many of the projects NSF reported as having a primary purpose to study EHS risks were part of its efforts to build a national research infrastructure capable of supporting a wide range of nanotechnology-related research. Specifically, NSF sponsors 16 Nanoscale Science and Engineering Centers, many of which devote a portion of their research efforts to EHS risk-related projects. In these cases, NSF apportioned a segment of the Center funding to the EHS category to account for this research. At NIOSH, both projects that we identified as not being primarily focused on studying EHS risks were focused on using nanotechnology to mitigate workplace risks, such as developing advanced sensors that incorporate nanotechnology to detect the presence of toxic gases in the workplace.

We found that the miscategorization of these 22 projects resulted largely from a reporting structure for nanotechnology research that does not easily allow agencies to recognize projects that use nanotechnology to improve the environment or enhance the detection of environmental contaminants, and from the limited guidance available to the agencies on how to consistently report EHS research. From fiscal years 2001 to 2004, the NSET subcommittee categorized federal research and development activities into nine categories, known as "grand challenges," that included one focused on "nanoscale processes for environmental improvement." Agencies funded and researchers initiated work on many of these 22 projects under the grand challenges categorization scheme. Starting in fiscal year 2005, NSET adopted a new categorization scheme for agencies to report their nanotechnology research. The new scheme, which was based on PCAs, eliminated the environmental improvement applications research category. Instead, agencies were asked to fund and report research designed to address or understand the risks associated with nanotechnology, as part of the societal dimensions PCA. In essence, the new scheme shifted the focus from applications-oriented research to research focused on the EHS implications of nanotechnology. However, under the new scheme, agencies no longer had a way to categorize environmentally focused research that had been initiated. As a result, NSF and NIOSH characterized these projects as EHS focused for lack of a more closely related category to place them in, according to program managers. Furthermore, neither NSET nor OMB provided agencies guidance on to how to apportion the dollars for a single project to more than one program component area, when appropriate. This is especially significant for broad, multiphase research projects, such as NSF's support to develop networks of research facilities with the capability to address a range of nanotechnology-related topics. Of the five agencies we reviewed, only NSF apportioned funds for a single project to more than one PCA.

Agencies Conduct Additional Research that Also Helps Advance Scientific Knowledge of Potential EHS Risks

In addition to research reported to the NNI as being primarily focused on the EHS risks of nanotechnology, some agencies conduct research that is not reflected in the EHS totals provided by the NNI either because they are not considered federal research agencies or because the primary purpose of the research was not to study EHS risks. For example, FDA, which does not have a specific research budget and does not generally track nanotechnology research spending, used a portion of its operating funds in fiscal years 2004 through 2007 to undertake 15 research projects to evaluate the potential health risks of nanomaterials in the products that it regulates. One such project focused on sunscreens that contain nanosized particles of titanium dioxide to better understand their potential to be absorbed into the body through the skin. Another project is designed to study the toxicological and immunological responses to nanoparticles that may be used in therapeutic drugs. A fundamental understanding of potential risks will help FDA develop guidance and make future regulatory decisions regarding the manufacture and use of FDA-regulated products using these materials, according to program managers.

In addition, as noted in the NNI's annual Supplement to the President's Budget, some agencies conduct research that results in information highly relevant to EHS risks but that was not primarily directed at understanding or addressing those risks and therefore is not captured in the EHS total. For example, NIH has research underway to develop drug delivery mechanisms that use nanotechnology. While the primary purpose of such research is to develop medical applications using nanotechnology, the research also provides information on how toxic the nanomaterials are, whether they accumulate in body tissues, and how they interact with the body at the cellular and molecular level. Agencies report funding data for such research in other PCAs, such as nanoscale devices and systems, rather than the EHS area. In addition, NIST conducts an array of nanotechnology research to accurately quantify the properties of nanomaterials and determine their size, shape, and chemical composition. This type of information is needed to understand and measure nanomaterials to ensure safe handling and protection against potential health or environmental hazards. However, NIST reports the funding data for such research under other PCAs such as instrumentation research, metrology, and standards

Processes to Identify and Prioritize Needed EHS Research Appear Reasonable and are Ongoing but a Comprehensive Research Strategy Has not Yet been Developed

Ongoing agency and NEHI working group efforts to identify and prioritize needed research related to the potential EHS risks of nanotechnology appear reasonable but have not as yet resulted in a comprehensive research strategy to guide EHS research across agencies. We found that the EHS risk research undertaken in fiscal year 2006 addressed a range of EHS topics, was generally consistent with both agency- and NEHIidentified research priorities, and focused on the priority needs within each category to varying degrees.

Agencies Have Identified their EHS Research Priorities

We determined that each agency's nanotechnology research priorities generally reflect its mission. For example, the priorities identified by FDA and CPSC are largely focused on the detection and safety of nanoparticles in the commercial products they regulate. On the other hand, EHS research priorities identified by NSF reflect its broader mission to advance science in general, and include a more diverse range of priorities, such as the safety and transport of nanomaterials in the environment, and the safety of nanomaterials in the workplace.

All eight agencies in our review have processes in place to identify and prioritize the research they need related to the potential EHS risks of nanotechnology. Most agencies have developed task forces or designated individuals to specifically consider nanotechnology issues and identify priorities, although the scope and exact purpose of these activities differ by agency. EPA, for example, formed a Nanomaterial Research Strategy Team to craft a long-term, focused plan to guide all of the agency's nanotechnology research. The strategy, which identifies EPA's research priorities around four key themes and seven scientific questions, is based in part on the agency's 2007 "Nanotechnology White Paper" that described scientific issues the agency should consider to help ensure safe development of nanotechnology and to understand the potential risks. At other agencies, particularly those that have little or no funding for nanotechnology research, specific individuals throughout the agency have been tasked to identify and prioritize EHS research needs. For example, CPSC has assigned individual staff responsible for different aspects related to consumer product safety, such as health scientists, to monitor trends in the use of nanomaterials in such products, which helps inform the agency's nanotechnology research priorities. Once identified, agencies communicate their EHS research priorities to the public and to the research community in a variety of ways, including publication in agency documents that specifically address nanotechnology issues, agency strategic plans or budget documents, agency Web sites, and presentations at public conferences or workshops.

NNI'S EFFORTS TO PRIORITIZE RESEARCH NEEDS ARE ONGOING

In addition to the efforts of individual agencies, the NSET subcommittee has engaged in an iterative prioritization process through its NEHI working group, although this process is not yet complete. First, in 2006, NEHI identified but did not prioritize five broad research categories and 75 more specific subcategories of needs where additional information was considered necessary to further evaluate the potential EHS risks of nanotechnology.[8] The report identified these five general research categories as (1) Instrumentation, Metrology, and Analytical Methods; (2) Nanomaterials and Human Health; (3) Nanomaterials and the Environment; (4) Health and Environmental Exposure Assessment; and (5) Risk Management Methods.[9] Second, following efforts to obtain public input on its 2006 report, NEHI released another report in August 2007,[10] in which it distilled the previous list of 75 unprioritized specific research needs into a set of five prioritized needs for each of the five general research categories.[11]

The NEHI working group has used these initial steps to identify the gaps between the needs and priorities it has identified and the research that agencies have underway. According to agency and NNI officials, once this gap analysis is complete, NEHI will formulate a long-term, overarching EHS research strategy. According to the August 2007 report, the proposed strategy will list NEHI's final research priorities, describe current federal EHS research, document the unmet needs, identify opportunities for interagency collaboration, and establish a process for periodic review. As envisioned, the EHS research strategy will serve as guidance for individual agencies as they develop their own research agendas and make funding decisions. NEHI plans to complete this overarching research strategy and issue a report in early 2008, according to NNI officials.

Agencies' and the NNI's Prioritization Processes Appear Reasonable

Despite the fact that a comprehensive research strategy for EHS research has yet to be finalized, the prioritization processes taking place within individual agencies and the NNI appear so far to be reasonable. Numerous agency officials said their agency's EHS research priorities were generally reflected both in the NEHI working group's 2006 research needs and 2007 research prioritization reports. Our comparison of agency nanotechnology priorities to the NNI's priorities corroborated their statements. Specifically, we found that all but one of the research priorities identified by individual agencies could be linked to one or more of the five general research categories. For example, OSHA's need for toxicity data and information related to exposure is reflected in the two general research categories of Health and Environmental Exposure Assessment and Nanomaterials and Human Health. According to agency officials, the alignment of agency priorities with the general research categories is particularly beneficial to the regulatory agencies, such as CPSC and OSHA, which do not conduct their own research, but rely instead on research agencies for data to inform their regulatory decisions.

In addition, we found that the primary purposes of agency projects underway in fiscal year 2006 were generally consistent with both agency priorities and the NEHI working group's research categories. Of these 97 projects, 43 were focused on Nanomaterials and Human Health, including all 18 of the projects funded by NIH. In addition, EPA, NIOSH, and NSF each undertook research for this general research category. EPA and NSF funded all 25 projects related to Nanomaterials and the Environment. These two general research categories accounted for 70 percent of all projects focused on EHS risks. Reflective of its relatively large EHS research budget and broad mission, NSF sponsored projects in each of the five general research categories. In contrast, all the research projects NIST sponsored were related to Instrumentation, Metrology, and Analytical Methods.

Agency research addressed each of the five general research categories and focused on the priority needs within each category to varying degrees. With the exception of the Human Health category, for which all specific needs were considered a top priority, 43 percent of projects addressed the two highest-priority needs in each category and 37 percent addressed the two lowest-priority needs. For example, 8 of the 11 projects in the Instrumentation, Metrology, and Analytic Methods category focused on the highest-priority need to "develop methods to detect nanomaterials in biological matrices, the environment, and the workplace." In contrast, of the 25 projects related to Nanomaterials and the Environment, 3 addressed the highest-priority need in the category—"understand the effects of engineered nanomaterials in individuals of a species and the applicability of testing schemes to measure effects"—and 11 addressed the fourth- ranked priority—"determine factors affecting the environmental transport of nanomaterials." Moreover, although the NEHI working group considered the five specific research priorities related to human health equally important, 19 of the 43 projects focused on a single priority— "research to determine the mechanisms of interaction between nanomaterials and the body at the molecular, cellular, and tissular levels." See table 3 for a summary of projects by agency and specific NEHI research priority.

Table 3. Research Primarily Focused on the Environmental, Health, and Safety Risks of Nanotechnology by Agency and Specific Nanotechnology Environmental and Health Implications Working Group Research Priority

	EPA	NIH	NIOSH	NIST	NSF	Total
Instrumentation, Metrology, and Analytical Methods	0	0	1	2	8	11
Develop methods to detect nanomaterials in biological matrices, the environment, and the workplace				1	7	8
Understand how chemical and physical modifications affect the properties of nanomaterials						0
Develop methods for standardizing assessment of particle size, size distribution, shape, structure, and surface area			1	1		2
Develop certified reference materials for chemical and physical characterization of nanomaterials						0
Develop methods to characterize a nanomaterial's spatio-chemical composition, purity, and heterogeneity					1	1
Nanomaterials and Human Health	4	18	10	0	11	43
1. Develop methods to quantify and characterize exposure to nanomaterials and characterize nanomaterials in biological matrices[a]	1	1	4		2	8
	EPA	NIH	NIOSH	NIST	NSF	Total
2. Understand the absorption and transport of nanomaterials throughout the human bodya	1	1			2	4
3. Establish the relationship between the properties of nanomaterials and uptake via the respiratory or digestive tracts or through the eyes or skin, and assess body burdena		5	3		1	9
4. Determine the mechanisms of interaction between nanomaterials and the body at the molecular, cellular, and tissular levelsa	1	10	3		5	19
5. Identify or develop appropriate in vitro and in vivo assays/models to predict in vivo human responses to nanomaterials exposurea	1	1			1	3
Nanomaterials and the Environment	5	0	0	0	20	25
1. Understand the effects of engineered nanomaterials in individuals of a species and the applicability of testing schemes to measure effects	1				2	3
2. Understand environmental exposures through identification of principle sources of exposure and exposure routes					1	1
3. Evaluate abiotic and ecosystem-wide effects					6	6
4. Determine factors affecting the environmental transport of nanomaterials	2				9	11

Table 3. (Continued)

	EPA	NIH	NIOSH	NIST	NSF	Total
5. Understand the transformation of nanomaterials under different environmental conditions	2				2	4
Health and Environmental Exposure Assessment	0	0	3	0	2	5
1. Characterize exposures among workers			2		1	3
2. Identify population groups and environments exposed to engineered nanoscale materials						0
3. Characterize exposure to the general population from industrial processes and industrial and consumer products containing nanomaterials						0
4. Characterize health of exposed populations and environments						0
5. Understand workplace processes and factors that determine exposure to nanomaterials			1		1	2
Risk Management Methods	1	0	7	0	5	13
1. Understand and develop best workplace practices, processes, and environmental exposure controls			4		2	6
2. Examine product or material life cycle to inform risk reduction decisions	1				1	2
3. Develop risk characterization information to determine and classify nanomaterials based on physical or chemical properties			1		2	3
4. Develop nanomaterial-use and safety-incident trend information to help focus risk management efforts						0
5. Develop specific risk communication approaches and materials			2			2
Total	10	18	21	2	46	97

Source: GAO analysis of agency data.
aPriorities given equal weight.

Despite the fact that the NEHI working group's priorities reflect individual agency priorities, some environmental and industry groups have called for a more top-down and directed approach to the NNI's prioritization efforts. In various congressional testimonies and in written comments on the

NEHI working group's draft reports, some groups have suggested that the NNI adopt a stronger, more autonomous role in setting the federal EHS research agenda. Some of these groups suggest that the NNI should have the authority to direct participating agencies to undertake research in specific EHS areas, its own budget authority, and the ability to shift EHS research dollars among the agencies. Proponents believe that this more centralized approach would help ensure that a cohesive EHS research strategy is implemented in a timely manner and that sufficient resources are dedicated to the highest-priority research.

However, such a strategy may not be consistent with historical approaches used to set federal research priorities and would be difficult to implement given how federal research currently is funded. Federal expenditures for research and development are regular budget items and are contained, along with other types of expenditures, within the budgets of more than 20 federal agencies. For some of these agencies, research is a major activity, and for others, it is a smaller part of a much larger set of programs. Centralizing nanotechnology research expenditures in a single existing agency or new agency would be difficult to achieve. In addition, agency officials we spoke with were generally satisfied with the current bottom-up, consensus-based approach. Moreover, they said the process has benefited from the in-depth expertise each agency has developed. For example, NIH played a large role in shaping the priorities for Nanomaterials and Human Health; NIST was heavily involved with Instrumentation, Metrology, and Analytical Methods; and NIOSH was a major contributor to the development of priorities for Health and Environmental Exposure Assessment. Some officials acknowledged that while the current approach has limitations, it benefits from the input of a broader range of stakeholders. According to one official, information bubbles up through the NNI structure and is utilized to inform and create a top-down vision, which then serves to guide agency funding decisions.

Coordination Processes Have Fostered Interagency Collaboration and Information-Sharing

Agency and NNI processes to coordinate research and other activities related to the potential EHS risks of nanotechnology have been generally effective, and have resulted in numerous interagency collaborations. In fact, all eight agencies in this review have collaborated on multiple occasions with other NEHI-member agencies on activities related to the EHS risks of nanotechnology. These EHS-related activities are consistent with the expressed goals of the larger NNI—to promote the integration of federal efforts through communication, coordination, and collaboration. The NEHI working group is at the center of this effort. Regular NEHI working group meetings, augmented by informal discussions, have provided a venue for agencies to exchange information on a variety of topics associated with EHS risks, including their respective research needs and opportunities for collaborations.

Interagency collaboration has taken many forms, including joint sponsorship of EHS-related research and workshops, the detailing of staff to other NEHI working group agencies, and various other general collaborations or memoranda of understanding. For example, FDA, NIST, and NIH's Nanotechnology Characterization Laboratory have initiated formal agreements to collaborate on research to characterize the physical and biological properties of nanomaterials used in cancer diagnosis and treatment.12 An FDA official said that this arrangement was developed primarily through discussions that occurred as a result of the agencies' participation in NEHI. Participation in NEHI has helped facilitate other types of interagency collaborations including a 2007 memorandum of understanding between EPA and NSF to create and fund research at a virtual Center for the Environmental Implications of Nanotechnology, detailing a CPSC toxicologist to a research laboratory office at EPA, and sponsoring international conferences on nanotechnology and occupational health by all NNI agencies, led by NIOSH, in 2005, 2006, and 2007. See table 4 for more examples of interagency collaboration.

Table 4. Examples of Agency Collaborations Related to Potential EHS Risks of Nanotechnology

General Collaborations
NIH's National Characterization Laboratory, in partnership with FDA and NIST, is developing characterization methods to evaluate nanomaterials intended for cancer treatments.
EPA and NSF signed a memorandum of understanding to create and fund research at a virtual Center for the Environmental Implications of Nanotechnology.
CPSC, FDA, NIH, and NIOSH have participated collaboratively on the Toxicological Evaluation of Nanoscale Materials program within the National Toxicology Program.
NIOSH and OSHA have collaborated to develop guidelines for working with engineered nanomaterials.
NIH and NIST have collaborated to characterize properties of nanoparticles commonly used in sunscreen lotions.
Staff from NIST and CPSC have been detailed to other NNI agencies.
Grant Solicitations
EPA, NIH, NIOSH, and NSF have issued interagency competitive grant announcements through EPA's Science to Achieve Results program to address various environmental and health implications of nanotechnology.
EPA, NIH, and NIOSH have developed an interagency Funding Opportunity Announcement to investigate the biocompatibility and toxicity of industrial nanomaterials in mammals.
EPA and NIOSH have funded research on the dispersion of nanoscale particulate aerosols.
Workshops
NIOSH and other NNI agencies have sponsored international conferences on nanotechnology and occupational health in 2005, 2006, and 2007.
NSF has facilitated meetings for NSF grantees on the EHS aspects of nanotechnology with participation from other NNI agencies.

Source: GAO.

Furthermore, the NEHI working group has adopted a number of practices GAO has previously identified as essential to helping enhance and sustain collaboration among federal agencies.[13] For example, NEHI's 2005 "Terms of Reference" clearly defined its purpose and objectives and delineated roles and responsibilities for group members. Furthermore, collaboration through multiagency grant announcements and jointly sponsored workshops has served as a mechanism to leverage limited resources to achieve increased knowledge about potential EHS risks. Despite the general effectiveness of its collaboration efforts, the NEHI working group has not yet completed an overarching strategy to help align the agencies' EHS research efforts. A completed strategy, combined with the results of the research needs prioritization process, also will serve as a means to monitor, evaluate, and report on the progress of meeting EHS research needs. In the meantime, the NNI's annual Supplements to the President's Budget have described the agencies' activities related to EHS issues, among other things, and provided a mechanism to reinforce agency accountability and performance.

Finally, all agency officials we spoke with expressed satisfaction with their agency's participation in the NEHI working group, specifically, the coordination and collaboration on EHS risk research and other activities that have occurred as a result of their participation. Many officials described NEHI as unique among interagency efforts in terms of its effectiveness. Given limited resources, the development of ongoing relationships between agencies with different missions, but compatible nanotechnology research goals, is particularly important. NIH officials commented that their agency's collaboration with NIST to develop standard reference materials for nanoparticles may not have occurred as readily had it not been for regular NEHI meetings and workshops. In addition, NEHI has effectively brought together research and regulatory agencies, which has enhanced planning and coordination. Many officials noted that participation in NEHI has frequently given regulators the opportunity to become aware of and involved with research projects at a very early point in their development, which has resulted in research that better suits the needs of regulatory agencies. Participation in NEHI is particularly important for agencies like CPSC, FDA, and OSHA that do not have dedicated budgets for nanotechnology research.

Many officials also cited the dedication of individual NEHI working group representatives, who participate in the working group in addition to their regular agency duties, as critical to the group's overall effectiveness. A number of the members has served on the body for several years, providing stability and continuity that contributes to a collegial and productive working atmosphere. In addition, because nanotechnology is relatively new with many unknowns, these officials said the agencies are excited about advancing knowledge about nanomaterials and contributing to the informational needs of both regulatory and research agencies. Furthermore, according to some officials, there is a shared sense among NEHI representatives of the need to apply lessons learned from the development of past technologies, such as genetically modified organisms, to help ensure the safe development and application of nanotechnology.

CONCLUSIONS

Nanotechnology is likely to affect many aspects of our daily lives in the future as novel drug delivery systems, improved energy storage capabilities, and stronger, lightweight materials are developed and made available to the public. However, for a technology that may become ubiquitous, it is essential to consider the potential risks of using nanotechnology in concert with its potential benefits. The first steps are to identify what is not known about the properties of nanomaterials and what must be known about how these materials interact with our bodies and our environment. The NNI, through its NEHI working group, has begun a process to identify and prioritize both the research needed to better understand potential EHS risks and the gaps between what research is underway and the highest-priority needs. Essential to this process is consistent, accurate, and complete information on the amount of agency research designed to address and understand EHS risks. However, this information is not currently available because the totals reported by the NNI include research that is more closely related to uses of nanotechnology, rather than the risks nanotechnology may pose. Furthermore, agencies currently have limited guidance on how to report projects with more than one research focus across program component areas, when appropriate. As a result, the inventory of projects designed to address these risks is inaccurate and cannot ensure that agencies direct their future research investments appropriately.

APPENDIX I. COMMENTS FROM THE OFFICE OF SCIENCE AND TECHNOLOGY POLICY

EXECUTIVE OFFICE OF THE PRESIDENT
OFFICE OF SCIENCE AND TECHNOLOGY POLICY
WASHINGTON, D.C. 20502

March 3, 2008

Ms. Anu Mittal
Director of Natural Resources and Environment
U.S. Government Accountability Office
441 G. Street N.W.
Washington, D.C. 20548

Dear Ms. Mittal:

Thank you for the opportunity to comment on the GAO's proposed report on Federally supported nanotechnology-related environmental, health, and safety research (GAO-08-402).

Generally, the Office of Science and Technology Policy (OSTP) agrees with the report's findings and is pleased that the GAO found interagency coordination and prioritization of nanotechnology EHS research to be effective and reasonable. OSTP is particularly pleased that the GAO noted one of the strengths of the government's interagency coordination lies in the knowledgeable and dedicated Federal employees that work in this area.

With respect to GAO's finding that a fraction of NNI-identified EHS research projects is not primarily focused on EHS risks, OSTP acknowledges that this budget cross-cut is open to reasonable, different interpretations. These illustrate the challenges associated with attempting to categorize fundamental, knowledge-creating research and to determine to what degree such research is related to EHS. Nonetheless, while GAO's presumption in the process recommendation—that more accurate reporting will better support program planning and policy development—is correct in principle, it does not follow that marginal differences in individual agency reporting negatively affect interagency coordination and prioritization or agency-level implementation of nanotechnology EHS research. This is borne out by GAO's own findings in this report.

OSTP, the Office of Management and Budget (OMB), and the National Nanotechnology Coordinating Office (NNCO) already provide extensive detailed guidance for reporting nanotechnology research, particularly for EHS—more guidance, in fact, than perhaps any other area of Federally-funded research and development. This guidance is based directly on program component area designations established in the National Nanotechnology Initiative Strategic Plan by the interagency Subcommittee on Nanoscale Science, Engineering, and Technology (NSET) of the National Science and Technology Council. As with all agency reporting, we fundamentally rely on the expert judgment of the program managers at the agency level.

Ms. Anu Mittal – Page 2

In the aggregate, the data that are currently reported provide sufficient and clear perspective and are just one of many key inputs to guide coordination, program planning and future directions. Therefore, OSTP does not agree that more detailed reporting guidance will result in better coordination of nanotechnology EHS research. However, given the GAO's findings, OSTP will discuss the manner in which agencies respond to current guidance at future NSET meetings.

Thank you for the opportunity to comment. OSTP appreciates GAO's efforts to fully understand the NNI and NSET interagency coordination process, and generally agrees with the report's overall assessments.

Sincerely,

John H. Marburger, III
Director

APPENDIX II. COMMENTS FROM THE DEPARTMENT OF HEALTH AND HUMAN SERVICES

DEPARTMENT OF HEALTH & HUMAN SERVICES

Office of the Assistant Secretary for Legislation

Washington, D.C. 20201

FEB 29 2008

Ms. Anu Mittal
Director, Natural Resources
and Environment
U.S. Government Accountability Office
Washington, DC 20548

Dear Ms. Mittal:

Enclosed are the Department's comments on the U.S. Government Accountability Office's (GAO) draft report entitled: Nanotechnology: Better Guidance Is Needed to Ensure Accurate Reporting of Federal Research Focused on Environmental, Health, and Safety Risks (GAO 08-402).

The Department appreciates the opportunity to review and comment on this draft report before its publication.

Sincerely,

for Vince Ventimiglia
Assistant Secretary for Legislation

GENERAL COMMENTS OF THE U.S. DEPARTMENT OF HEALTH AND HUMAN SERVICES (HHS) ON THE U.S. GOVENRMENT ACCOUNTABILITY OFFICE'S DRAFT (GAO) DRAFT REPORT ENTITLED: "NANOTECHNOLOGY: BETTER GUIDANCE IS NEEDED TO ENSURE ACCURATE REPORTING OF REDERAL RESEARCH FOCUSED ON ENVIRONMENTAL, HEALTH, AND SAFETY RISKS "(GAO-08-402)

The report clearly addresses the three charges that GAO was given, but in finding that there were some misclassifications of "applications" projects as "implications," the report could have focused more on the adequacy of funding for research on environmental and occupational health and safety implications. Additionally, the report does not distinguish funding derived from NNI allocations and funding resulting if an agency reprograms non NNI internal funds.

The report does not discuss the activities of OSHA regarding occupational health risks.

APPENDIX III. COMMENTS FROM THE NATIONAL INSTITUTE OF STANDARDS AND TECHNOLOGY

UNITED STATES DEPARTMENT OF COMMERCE
National Institute of Standards and Technology
Gaithersburg, Maryland 20899-0001
OFFICE OF THE DIRECTOR

FEB 2 1 2008

MEMORANDUM FOR Anu Mittal
Director, Natural Resources and Environment
Government Accountability Office

From: James M. Turner, Ph.D.
Acting Director

Subject: Comments on Government Accountability Office (GAO) Draft Report Entitled "NANOTECHNOLOGY: Better Guidance is Needed to Ensure Accurate Reporting of Federal Research Focused on Environmental, Health, and Safety Risks" (GAO-08-402)

This is in response to your draft report dated March 2008 entitled "NANOTECHNOLOGY: Better Guidance is Needed to Ensure Accurate Reporting of Federal Research Focused on Environmental, Health, and Safety Risks." Thank you for the opportunity to review and comment on this draft.

Overall, the draft report is fair and balanced. The report does not, however, explicitly incorporate changes requested by the National Institute of Standards and Technology (NIST) to elucidate the relatively large size of nanotechnology Environmental Health and Safety (EHS) - *enabling* NIST research compared to the small size of explicitly nanotechnology EHS - *focused* Program Component Area (PCA) 7A research.

The report accurately notes that "Some agencies conduct research that results in information highly relevant to EHS risks but that this research was not primarily directed at understanding, or addressing, those risks and therefore is not captured in the EHS total." Further, the report gives two detailed examples of research resulting in information highly relevant to EHS risks, and included NIST in these examples. I appreciate the recognition of this situation and agree with the findings.

I am committed to ensuring that NIST continues to provide technically sound and timely standards support to the National Nanotechnology Initiative (NNI).

We are looking forward to receiving your final report. Please contact Steve Willett on (301) 975-8707 should you have any questions regarding this response.

NIST

REFERENCES

[1] The creation of the NNI formalized an existing interagency dialogue on nanotechnology that began in 1998.

[2] For purposes of this report we use the term "research agency" to mean an agency whose primary mission is to conduct or facilitate scientific research, and the term "regulatory agency" to mean an agency whose primary mission is to administer regulatory programs related to environment, human health, and safety, and which may have a role in regulating products containing nanomaterials. A small number of agencies carry out both functions;we will refer to these by their primary mission.

[3] As of December 2007, a total of four working groups exist within the NSET subcommittee: (1) Global Issues in Nanotechnology; (2) Nanotechnology Environmental and Health Implications; (3) Nanomanufacturing, Industry Liaison, and Innovation; and (4) Nanotechnology Public Engagement and Communications.

[4] Pub. L. No. 108-153 (2003).

[5] The nine grand challenges were as follows: nanostructured materials by design; manufacturing at the nanoscale; chemical-biological-radiological-explosive detection, and protection; nanoscale instrumentation and metrology; nano-electronics, -photonics, and - magnetics; healthcare, therapeutics, and diagnostics; efficient energy conversion and storage; microcraft and robotics; and nanoscale processes for environmental improvement.

[6] The NNI's four goals are to (1) maintain a world-class research and development program aimed at realizing the full potential of nanotechnology; (2) facilitate transfer of new technologies into products for economic growth, jobs, and other public benefit; (3) develop educational resources, a skilled workforce, and the supporting infrastructure and tools to advance nanotechnology; and (4) support responsible development of nanotechnology.

[7] The seven program component areas are fundamental nanoscale phenomena and processes; nanomaterials; nanoscale devices and systems; instrumentation research, metrology, and standards for nanotechnology; nanomanufacturing; major research facilities and instrumentation acquisition; and societal dimensions.

[8] NSTC, Committee on Technology, Subcommittee on Nanoscale Science, Engineering, and Technology, "Environmental, Health, and Safety Research Needs for Engineered Nanoscale Materials," Sept. 20, 2006.

[9] The Health and Environmental Exposure Assessment category was initially named Health and Environmental Surveillance.

[10] NSTC, Committee on Technology, Subcommittee on Nanoscale Science, Engineering, and Technology, NEHI Working Group, "Prioritization of Environmental, Health, and Safety Research Needs for Engineered Nanoscale Materials: An Interim Document for Public Comment," Aug. 16, 2007.

[11] The five specific needs for the Nanomaterials and Human Health category were all afforded the same top priority

[12] The Nanotechnology Characterization Laboratory is part of NIH's National Cancer Institute

[13] GAO, Results-Oriented Government: Practices That Can Help Enhance and Sustain Collaboration among Federal Agencies, GAO-06-15 (Washington, D.C.: Oct. 21, 2005).

In: Nanotechnology in the USA: Developments, Policies and Issues ISBN 978-1-60692-800-4
Editor: Carl H. Jennings

Chapter 5

NANOTECHNOLOGY: ACCURACY OF DATA ON FEDERALLY FUNDED ENVIRONMENTAL, HEALTH, AND SAFETY RESEARCH COULD BE IMPROVED*

Statement of Robert A. Robinson

WHY UNITED STATES GOVERNMENT ACCOUNTABILITY OFFICE DID THIS STUDY

In March 2008, GAO issued a report entitled Nanotechnology: Better Guidance Is Needed to Ensure Accurate Reporting of Federal Research Focused on Environmental, Health, and Safety Risks (GAO-08-402). In this report, GAO reviewed the National Nanotechnology Initiative (NNI), a multiagency effort administered by the Office of Science and Technology Policy (OSTP). The NNI coordinates the nanotechnology-related activities of 25 federal agencies that fund nanoscale research or have a stake in the results. A key research area funded by some agencies related to studying the potential environmental, health, and safety (EHS) risks that may result from exposure to nanoscale materials. For this testimony statement, GAO was asked to summarize the findings of its March 2008 report, focusing on (1) the extent to which selected agencies conducted EHS research in fiscal year 2006; (2) the reasonableness of the agencies' and the NNI's processes to identify and prioritize EHS research; and (3) the effectiveness of the agencies' and the NNI's process to coordinate EHS research.

* This is an edited, excerpted and augmented edition of a United States Government Accountability Office Report GAO-08-709T, Dated , April 24, 2008.

What United States Government Accountability Office Reccomends

In its March 2008 report, GAO recommended better guidance to improve the accuracy of data reported by the NNI. Although OSTP asserted that it provides extensive guidance, it agreed to review how the agencies respond to the current guidance. GAO is making no new recommendations in this statement.

What United States Government Accountability Office Found

In fiscal year 2006, federal agencies devoted $37.7 million—or 3 percent of the $1.3 billion total nanotechnology research funding—to research that was primarily focused on the EHS risks of nanotechnology, according to the NNI. However, about 20 percent of this total cannot actually be attributed to this purpose. GAO found that 22 of the 119 projects identified as EHS in fiscal year 2006 were not primarily related to understanding the extent to which nanotechnology may pose an EHS risk. Instead, many of these projects were focused on how to use nanotechnology to remediate environmental damage or detect hazards not related to nanotechnology. GAO determined that this mischaracterization is rooted in the current reporting structure that does not allow these types of projects to be easily categorized and the lack of guidance for agencies on how to apportion research funding across multiple topics, when appropriate. In addition to the EHS funding reported by the NNI, federal agencies conduct other research that is not captured in the EHS totals. This research was not captured by the NNI because either the research was funded by an agency not considered to be a research agency or because the primary purpose of the research was not to study EHS risks.

Federal agencies and the NNI, at the time of GAO's review, were in the process of identifying and prioritizing EHS risk research needs and the overall process they were using appeared reasonable. For example, identification and prioritization of EHS research needs was being done by the agencies and the NNI collaboratively. The NNI also was engaged in an iterative prioritization effort through its Nanotechnology Environmental and Health Implications (NEHI) working group. Through this process, NEHI identified five general research categories as a priority for federally funded research. GAO found that most of the research projects that were underway in fiscal year 2006 were generally consistent with agency and NEHI priorities. NEHI released its new EHS research strategy on February 13, 2008.

Agency and NNI processes to coordinate activities related to potential EHS risks of nanotechnology have been generally effective. The NEHI working group has convened frequent meetings that have helped agencies identify opportunities to collaborate on EHS risk issues, such as joint sponsorship of research and workshops to advance knowledge and facilitate information- sharing among the agencies. NEHI also has incorporated several practices that GAO has previously identified as key to enhancing and sustaining interagency collaborative efforts, such as defining a common outcome and leveraging resources. Finally, all agency officials GAO spoke with expressed satisfaction with the coordination and collaboration on EHS risk research that has occurred through NEHI. They cited several factors they believe contribute to the group's effectiveness, including the stability of the working group membership and the expertise and dedication of its members. Furthermore, according to these officials, this stability, combined with common research needs and general excitement about the new science, has resulted in a collegial, productive working environment.

Mr. Chairman and Members of the Subcommittee:

I am pleased to be here today to participate in your hearing on the future direction of the National Nanotechnology Initiative (NNI). As you know, the NNI was established in 2001 as a federal, multiagency effort intended to accelerate the discovery, development, and deployment of nanoscale science, engineering, and technology to achieve economic benefits, enhance the quality of life, and promote national security. One of the key roles of the NNI is to coordinate the nanotechnology-related activities of 25 federal agencies. These agencies include both those that fund nanoscale research as well as those that have a stake in the outcome of this research, such as agencies that regulate products containing nanomaterials. While the NNI is designed to facilitate intergovernmental cooperation and identify goals and priorities for nanotechnology research, it is not a research program. It has no funding or authority to dictate the nanotechnology research agenda for participating agencies or to ensure that adequate resources are available to achieve specific goals. Instead, participating agencies develop and fund their own nanotechnology research agendas, and in fiscal year 2006, 13 of the 25 agencies participating in the NNI allocated a total of about $1.3 billion from their appropriated budgets to nanotechnology research and development activities. Of this total in fiscal year 2006, the NNI reported that $37.7 million (or about 3 percent of the total) was used to fund research to study the potential environmental, health, and safety (EHS) risks that might result from exposure during the manufacture, use, and disposal or recycle of nanoscale materials. As you know, while the use of nanoscale materials holds much promise, the small size and unique properties of these materials raise questions about their potential EHS risks, and research is needed to fill current gaps in scientific information about their risks.

At the request of the full committee and members of the Congressional Nanotechnology Caucus, we just completed a report that is being released today on the NNI's and federal agencies' efforts to study the potential environmental, health, and safety risks of nanotechnology.1 My testimony is based on the findings of this review and will cover the following three areas: (1) the extent to which selected research and regulatory agencies conducted research in fiscal year 2006 that primarily was focused on the potential EHS risks of nanotechnology; (2) the reasonableness of the processes that agencies and the NNI use to identify and prioritize federal research on the potential EHS risks of nanotechnology; and (3) the effectiveness of the processes that agencies and the NNI use to coordinate their research. For our review, we collected data from five federal agencies that provided 96 percent of fiscal year 2006 funding for EHS research—the Environmental Protection Agency (EPA), the National Institutes of Health (NIH), the National Institute for Occupational Safety and Health (NIOSH), the National Institute of Standards and Technology (NIST), and the National Science Foundation (NSF). We also contacted three regulatory agencies—the U.S. Consumer Product Safety Commission (CPSC), the Food and Drug Administration (FDA), and the Occupational Safety and Health Administration (OSHA)—that do not have specific research budgets to determine whether they conducted any research on their own relative to potential EHS risks. We conducted this performance audit from June 2007 to February 2008 in accordance with generally accepted government auditing standards. Those standards require that we plan and perform the audit to obtain sufficient, appropriate evidence to provide a reasonable basis for our finding and conclusions based on our audit objectives. We believe that the evidence obtained provides a reasonable basis for our findings and conclusions based on our audit objectives.

In summary we found the following:

About 20 percent of the over $37 million in fiscal year 2006 research expenditures that the NNI reported as being primarily focused on the EHS risks of nanotechnology cannot actually be attributed to this purpose. We found that 22 of the 119 projects identified as EHS-related by EPA, NIH, NIOSH, NIST, and NSF in fiscal year 2006 were not primarily related to understanding the extent to which nanotechnology poses an EHS risk. These 22 projects, funded by NSF and NIOSH, accounted for about $7 million of the $37 million that the NNI reported as being primarily focused on EHS risks. The focus of many of these projects was to explore how nanotechnology could be used to remediate environmental damage or to detect a variety of hazards unrelated to nanotechnology. We determined that this mischaracterization was the result of the current reporting structure that does not allow these types of projects to be easily categorized in another more appropriate category, and also the lack of guidance for agencies on how to apportion research funding across multiple topics, when appropriate. To address this issue, we recommended that the Office of Science and Technology Policy (OSTP), in consultation with the NNI and the Office of Management and Budget (OMB), provide better guidance to agencies regarding how to report research that is primarily focused on understanding or addressing the EHS risks of nanotechnology. In commenting on this report, OSTP asserted that it already provides extensive guidance, but it agreed to review the manner in which agencies respond to the current guidance.

- In addition to the EHS funding totals reported by the NNI, we found that federal agencies conduct other EHS research that is not captured in the NNI totals. This research was not captured by the NNI because either the research was funded by an agency not generally considered to be a research agency or because the primary purpose of the research was not to study EHS risks. Because the agencies that conduct this research do not systematically track it as EHS-related research, we could not establish the exact amount of federal funding that is being devoted to this additional EHS research.
- Federal agencies and the NNI were, at the time of our review, in the process of identifying and prioritizing EHS risk research needs; overall, we believe that the process they were using was reasonable. For example, identification and prioritization of EHS research needs was being done by the agencies and the NNI collaboratively. The NNI was also engaged in an iterative prioritization effort through its Nanotechnology Environmental and Health Implications (NEHI) working group. As a result of this effort, NEHI had identified five general research categories that should be the focus of federal research efforts and five specific research priorities under each general category. Our analysis of the 97 research projects that were underway in fiscal year 2006 that were primarily related to studying EHS risks

found that the focus of these projects was generally consistent with agency priorities as well as NEHI's five general research categories. However, we did find that, while agency funded research addressed each of the five general research categories, it focused on the priority needs within each category to varying degrees. As our report was in production, NEHI released a new EHS research strategy on February 13, 2008, which is intended to provide a framework to help ensure that the highest priority EHS research needs are met.

- Agency and NNI processes to coordinate activities related to potential EHS risks of nanotechnology have been generally effective. The NEHI working group has convened frequent meetings that have helped agencies identify opportunities to collaborate on EHS risk issues, such as joint sponsorship of research and workshops to advance knowledge and facilitate information-sharing among the agencies. These types of exchanges, according to most agency officials we spoke with, have helped advance knowledge and facilitated information-sharing among the agencies. In addition, NEHI has incorporated several practices that we have previously identified as key to enhancing and sustaining interagency collaborative efforts, such as defining a common outcome and leveraging resources, but it had not, at the time of our review, completed its overarching strategy to help better align agencies' EHS research efforts. Finally, all agency officials we spoke with expressed satisfaction with the coordination and collaboration on EHS risk research that has occurred through NEHI. They cited several factors they believe contribute to the group's effectiveness, including the stability of the working group membership and the expertise and dedication of its members. Furthermore, according to these officials, this stability, combined with common research needs and general excitement about the new science, has resulted in a collegial, productive working environment

BACKGROUND

Nanotechnology encompasses a wide range of innovations based on the understanding and control of matter at the scale of nanometers—the equivalent of one-billionth of a meter. To illustrate, a sheet of paper is about 100,000 nanometers thick and a human hair is about 80,000 nanometers wide. At the nanoscale level, materials may exhibit electrical, biological, and other properties that differ significantly from the properties the same materials exhibit at a larger scale. Exploiting these differences in nanoscale materials has led to a range of commercial uses and holds the promise for innovations in virtually every industry from aerospace and energy to health care and agriculture. In 2006, an estimated $50 billion in products worldwide incorporated nanotechnology and this figure has been projected to grow to $2.6 trillion by 2014. One research institute estimates that over 500 consumer products already available to consumers may contain nanoscale materials.

The National Nanotechnology Initiative (NNI) was established in 2001 as a federal, multiagency effort intended to accelerate the discovery, development, and deployment of nanoscale science, engineering, and technology to achieve economic benefits, enhance the quality of life, and promote national security. Management of the NNI falls under the purview of the National Science and Technology Council (NSTC) that coordinates science and technology policy across the federal government. The NSTC is managed by the Director of the Office of Science and Technology Policy (OSTP), who also serves as the Science Advisor to the President. The NSTC's Committee on Technology established the Nanoscale Science, Engineering, and Technology (NSET) subcommittee to help coordinate, plan, and implement the NNI's activities across participating agencies. In 2003, the NSET subcommittee further established a Nanotechnology.

Environmental and Health Implications (NEHI) working group.[2] The purpose of the NEHI working group, composed of representatives from 16 research and regulatory agencies, is to, among other things, coordinate agency efforts related to EHS risks of nanotechnology. Similar to the NNI, the NEHI working group has no authority to mandate research priorities or to ensure that agencies adequately fund particular research.

In December 2003, Congress enacted legislation to establish a National Nanotechnology Program to coordinate federal nanotechnology research and development.[3]Among other things, the act directs the NSTC to establish goals and priorities for the program and to set up program component areas that reflect those goals and priorities. To implement these requirements, the NSTC has established a process to categorize research projects and activities undertaken by the various federal agencies into seven areas. Six of the seven focus on the discovery, development, and deployment of nanotechnology, while the seventh relates to the societal dimensions of nanotechnology that include issues such as the EHS risks of nanotechnology.

As part of the annual federal budget process, agencies also report their research funding for each area to OMB. The NNI's annual Supplement to the President's Budget, prepared by the NSTC, includes EHS research figures from the agencies and a general description of the research conducted by the agencies in each of the areas. For reporting purposes, the NSET subcommittee has defined EHS research as efforts whose primary purpose is to understand and address potential risks to health and to the environment posed by nanotechnology. Eight of the 13 agencies that funded nanotechnology research in fiscal year 2006 reported having devoted some of those resources to research that had a primary focus on potential EHS risks.

Under the NNI, each agency funds research and development projects that support its own mission as well as the NNI's goals. While agencies share information on their nanotechnology-related research goals with the NSET subcommittee and NEHI working group, each agency retains control over its decisions on the specific projects to fund. While the NNI was designed to facilitate intergovernmental cooperation and identify goals and priorities for nanotechnology research, it is not a research program. It has no funding or authority to dictate the nanotechnology research agenda for participating agencies.

The NNI used its fiscal year 2000 strategic plan and its subsequent updates to delineate a strategy to support long-term nanoscale research and development, among other things. A key component of the 2000 plan was the identification of nine specific research and development areas— known as "grand challenges"—that highlighted federal research on applications of nanotechnology with the potential to realize significant economic, governmental, and societal benefits.

In 2004, the NNI updated its strategic plan and described its goals as well as the investment strategy by which those goals were to be achieved. Consistent with the 21st Century Nanotechnology Research and Development Act, the NNI reorganized its major subject categories of research and development investment into program component areas (PCA) that cut across the interests and needs of the participating agencies. These seven areas replaced the nine grand challenges that the agencies had used to categorize their nanotechnology research. Six of the areas focus on the discovery, development, and deployment of nanotechnology. The seventh, societal dimensions, consists of two topics—research on environmental, health, and safety; and education and research on ethical, legal, and other societal aspects of nanotechnology.

PCAs are intended to provide a means by which the NSET subcommittee, OSTP, OMB, Congress, and others may be informed of the relative federal investment in these key areas. PCAs also provide a structure by which the agencies that fund research can better direct and coordinate their activities. In response to increased concerns about the potential EHS risks of nanotechnology, the NSET subcommittee and the agencies agreed in fiscal year 2005 to separately report their research funding for each of the two components of the societal dimensions PCA. The December 2007 update of the NNI's strategic plan reaffirmed the program's goals, identified steps to accomplish those goals, and formally divided the societal dimensions PCA into two PCAs—"environment, health, and safety" and "education and societal dimensions."

Beginning with the development of the fiscal year 2005 federal budget, agencies have worked with OMB to identify funding for nanoscale research that would be reflected in the NNI's annual Supplement to the President's Budget. OMB analysts reviewed aggregated, rather than project-level, data on research funding for each PCA to help ensure consistent reporting across the agencies. Agencies also relied on definitions of the PCAs developed by the NSET subcommittee to determine the appropriate area in which to report research funding. Neither NSET nor OMB provided guidance on whether or how to apportion funding for a single research project to more than one PCA, if appropriate. However, representatives from both NSET and OMB stressed that the agencies were not to report each research dollar more than once.

Almost One-Fifth of Reported EHS Research Projects Were Not Primarily Focused on Studying the EHS Risks of Nanotechnology

About 18 percent of the total research dollars reported by the agencies as being primarily focused on the study of nanotechnology-related EHS risks in fiscal year 2006 cannot actually be attributed to this purpose. Specifically, we found that 22 of the 119 projects funded by five federal agencies were not primarily related to studying EHS risks. These 22 projects accounted for about $7 million of the total that the NNI reported as supporting research primarily focused on EHS risks. Almost all of these projects—20 out of 22—were funded by NSF, with the two additional projects funded by NIOSH. We found that the primary purpose of many of these 22 projects was to explore ways to use nanotechnology to remediate environmental damage or to identify environmental, chemical, or biological hazards not related to nanotechnology. For example, some NSF- funded research explored the use of nanotechnology to improve water or gaseous filtration systems. Table 1 shows our analysis of the nanotechnology research projects reported as being primarily focused on EHS risks.

Table 1. GAO Analysis of the Number and Dollar Value of Nanotechnology Research Projects Reported by Selected Agencies as Being Primarily Focused on Environmental, Health, and Safety Risks, Fiscal Year 2006

(Dollars in millions)

Projects reported by agencies as being primarily focused on EHS			Projects determined by GAO to be primarily focused on EHS		Projects determined by GAO not to be primarily focused on EHS	
Agency	Number	Dollar Value[a]	Number	Dollar Value	Number	Dollar value
EPA	10	$3.6	10	$3.6	0	$0
NIH	18	$5.6	18	$5.6	0	$0
NIOSH	23	$4.3	21	$4.2	2	$0.1
NIST	2	$2.4	2	$2.4	0	$0
NSF	66	$21.1	46	$14.7	20	$6.4
Total	119	$37	97	$30.5	22	$6.5

Source: GAO analysis of agency obligations data.

a. Figures differ slightly from those reported by the NNI in the Supplement to the President's FY2008 Budget due to rounding error or modifications made to the project-level data after they were reported by agencies to the NNI.

We found that the miscategorization of these 22 projects resulted largely from a reporting structure for nanotechnology research that does not easily allow agencies to recognize projects that use nanotechnology to improve the environment or enhance the detection of environmental contaminants, and from the limited guidance available to the agencies on how to consistently report EHS research. From fiscal years 2001 to 2004, the NSET subcommittee categorized federal research and development activities into nine categories, known as "grand challenges," that included one focused on "nanoscale processes for environmental improvement." Agencies initiated work on many of these 22 projects under the grand challenges categorization scheme. Starting in fiscal year 2005, NSET adopted a new categorization scheme, based on PCAs, for agencies to report their nanotechnology research. The new scheme eliminated the research category of environmental improvement applications and asked agencies to report research designed to address or understand the risks associated with nanotechnology as part of the societal dimensions PCA.

The new scheme shifted the focus from applications-oriented research to research focused on the EHS implications of nanotechnology. However, the new scheme had no way for agencies to categorize environmentally focused research that was underway. As a result, NSF and NIOSH characterized these projects as EHS focused for lack of a more closely related category to place them in, according to program managers. Furthermore, neither NSET nor OMB provided agencies guidance on how to apportion the dollars for a single project to more than one program component area, when appropriate. This is especially significant for broad, multiphase research projects, such as NSF's support to develop networks of research facilities. Of the five agencies we reviewed, only NSF apportioned funds for a single project to more than one PCA.

In addition to research reported to the NNI as being primarily focused on the EHS risks of nanotechnology, some agencies conduct research that is not reflected in the EHS totals provided by the NNI either because they are not considered federal research agencies or because the primary purpose of the research was not to study EHS risks. For example, some agencies conduct research that results in information highly relevant to EHS risks but that was not primarily directed at understanding or addressing those risks and therefore is not captured in the EHS total. This type of research provides information that is needed to understand and measure nanomaterials to ensure safe handling and protection against potential health or environmental hazards; however, such research is captured under other PCAs, such as instrumentation, metrology, and standards. Because the agencies that conduct this research do not systematically track it as EHS-related, we could not establish the exact amount of federal funding that is being devoted to this additional EHS research.

Processes to Identify and Prioritize Needed EHS Research Appear Reasonable and Are Ongoing but a Comprehensive Research Strategy Has not Yet been Developed

All eight agencies in our review have processes in place to identify and prioritize the research they need related to the potential EHS risks of nanotechnology. Most agencies have developed task forces or designated individuals to specifically consider nanotechnology issues and identify priorities, although the scope and exact purpose of these activities differ by agency. Once identified, agencies communicate their EHS research priorities to the public and to the research community in a variety of ways, including publication in agency documents that specifically address nanotechnology issues, agency strategic plans or budget documents, agency Web sites, and presentations at public conferences or workshops. We determined that each agency's nanotechnology research priorities generally reflected its mission. For example, the priorities identified by FDA and CPSC are largely focused on the detection and safety of nanoparticles in the commercial products they regulate. On the other hand, EHS research priorities identified by NSF reflect its broader mission to advance science in general, and include a more diverse range of priorities, such as the safety and transport of nanomaterials in the environment, and the safety of nanomaterials in the workplace.

In addition to the efforts of individual agencies, the NSET subcommittee has engaged in an iterative prioritization process through its NEHI working group. Beginning in 2006, NEHI identified but did not prioritize five broad research categories and 75 more specific subcategories of needs where additional information was considered necessary to further evaluate the potential EHS risks of nanotechnology. NEHI obtained public input on its 2006 report and released another report in August 2007, in which it distilled the previous list of 75 unprioritized specific research needs into a set of five prioritized needs for each of the five general research categories. The NEHI working group has used these initial steps to identify the gaps between the needs and priorities it has identified and the research that agencies have underway. NEHI issued a report summarizing the results of this analysis in February 2008.

Although a comprehensive research strategy for EHS research had not been finalized at the time of our review, the prioritization processes taking place within individual agencies and the NNI appeared to be reasonable. Numerous agency officials said their agency's EHS research priorities were generally reflected both in the NEHI working group's 2006 research needs and 2007 research prioritization reports. Our comparison of agency nanotechnology priorities to the NNI's priorities corroborated these statements. Specifically, we found that all but one of the research priorities identified by individual agencies could be linked to one or more of the five general research categories. According to agency officials, the alignment of agency priorities with the general research categories is particularly beneficial to the regulatory agencies, such as CPSC and OSHA, which do not conduct their own research, but rely instead on research agencies for data to inform their regulatory decisions.

In addition, we found that the primary purposes of agency projects underway in fiscal year 2006 were generally consistent with both agency priorities and the NEHI working group's research categories. Of these 97 projects, 43 were focused on Nanomaterials and Human Health, including all 18 of the projects funded by NIH. EPA and NSF funded all 25 projects related to Nanomaterials and the Environment. These two general research categories accounted for 70 percent of all projects focused on EHS risks.

Furthermore, we determined that, while agency-funded research addressed each of the five general research categories, it focused on the priority needs within each category to varying degrees. Specifically, we found that the two highest-priority needs in each category were addressed only slightly more frequently than the two lowest-priority needs.

Moreover, although the NEHI working group considered the five specific research priorities related to human health equally important, 19 of the 43 projects focused on a single priority—"research to determine the mechanisms of interaction between nanomaterials and the body at the molecular, cellular, and tissular levels." Table 2 shows a summary of projects by agency and specific NEHI research priority.

Table 2. Research Primarily Focused on the Environmental, Health, and Safety Risks of Nanotechnology by Agency and Specific Nanotechnology Environmental and Health Implications Working Group Research Priority

	EPA	NIH	NIOSH	NIST	NSF	Total
Instrumentation, Metrology, and Analytical Methods	0	0	1	2	8	11
1. Develop methods to detect nanomaterials in biological matrices, the environment, and the workplace				1	7	8
2. Understand how chemical and physical modifications affect the properties of nanomaterials						0
3. Develop methods for standardizing assessment of particle size, size distribution, shape, structure, and			1	1		2
4. Develop certified reference materials for chemical and physical characterization of nanomaterials						0
5. Develop methods to characterize a nanomaterial's spatio-chemical composition, purity, and heterogeneity					1	1
Nanomaterials and Human Health	4	18	10	0	11	43
1. Develop methods to quantify and characterize exposure to nanomaterials and characterize nanomaterials in biological matrices[a]	1	1	4		2	8
2. Understand the absorption and transport of nanomaterials throughout the human body[a]	1	1			2	4
3. Establish the relationship between the properties of nanomaterials and uptake via the respiratory or digestive tracts or through the eyes or skin, and assess body burden[a]		5	3		1	9

	EPA	NIH	NIOSH	NIST	NSF	Total
4. Determine the mechanisms of interaction between nanomaterials and the body at the molecular, cellular, and tissular levels[a]	1	10	3		5	19
5. Identify or develop appropriate in vitro and in vivo assays/models to predict in vivo human responses to nanomaterials exposure[a]	1	1			1	3
Nanomaterials and the Environment	5	0	0	0	20	25
1. Understand the effects of engineered nanomaterials in individuals of a species and the applicability of testing schemes to measure effects	1				2	3
2. Understand environmental exposures through identification of principle sources of exposure and exposure routes					1	1
3. Evaluate abiotic and ecosystem-wide effects					6	6
4. Determine factors affecting the environmental transport of nanomaterials	2				9	11
5. Understand the transformation of nanomaterials under different environmental conditions	2				2	4
	EPA	**NIH**	**NIOSH**	**NIST**	**NSF**	**Total**
Health and Environmental Exposure Assessment	0	0	3	0	2	5
1. Characterize exposures among workers			2		1	3
2. Identify population groups and environments exposed to engineered nanoscale materials						0
3. Characterize exposure to the general population from industrial processes and industrial and consumer products containing nanomaterials						0
4. Characterize health of exposed populations and environments						0
5. Understand workplace processes and factors that determine exposure to nanomaterials			1		1	2
Risk Management Methods	1	0	7	0	5	13
1. Understand and develop best workplace practices, processes, and environmental exposure controls			4		2	6
2. Examine product or material life cycle to inform risk reduction decisions	1				1	2
3. Develop risk characterization information to determine and classify nanomaterials based on physical or chemical			1		2	3
4. Develop nanomaterial-use and safety-incident trend information to help focus risk management efforts						0
5. Develop specific risk communication approaches and materials			2			2
Total	10	18	21	2	46	97

Source: GAO analysis of agency data.

aPriorities given equal weight.

Coordination Processes Have Fostered Interagency Collaboration and Information-Sharing

Agency and NNI processes to coordinate research and other activities related to the potential EHS risks of nanotechnology have been generally effective, and have resulted in numerous interagency collaborations. All eight agencies in our review have collaborated on multiple occasions with other NEHI-member agencies on activities related to the EHS risks of nanotechnology. These EHS-related activities are consistent with the expressed goals of the larger NNI—to promote the integration of federal efforts through communication, coordination, and collaboration. The NEHI working group is at the center of this effort.

We found that regular NEHI working group meetings, augmented by informal discussions, have provided a venue for agencies to exchange information on a variety of topics associated with EHS risks, including their respective research needs and opportunities for collaborations. Interagency collaboration has taken many forms, including joint sponsorship of EHS-related research and workshops, the detailing of staff to other NEHI working group agencies, and various other general collaborations or memoranda of understanding.

Furthermore, the NEHI working group has adopted a number of practices GAO has previously identified as essential to helping enhance and sustain collaboration among federal agencies.[4] For example, in 2005 NEHI clearly defined its purpose and objectives and delineated roles and responsibilities for group members. Furthermore, collaboration through multiagency grant announcements and jointly sponsored workshops has served as a mechanism to leverage limited resources to achieve increased knowledge about potential EHS risks.

Finally, all agency officials we spoke with expressed satisfaction with their agency's participation in the NEHI working group, specifically, the coordination and collaboration on EHS risk research and other activities that have occurred as a result of their participation. Many officials described NEHI as unique among interagency efforts in terms of its effectiveness. Given limited resources, the development of ongoing relationships between agencies with different missions, but compatible nanotechnology research goals, is particularly important. NIH officials commented that their agency's collaboration with NIST to develop standard reference materials for nanoparticles may not have occurred as readily had it not been for regular NEHI meetings and workshops. In addition, NEHI has effectively brought together research and regulatory agencies, which has enhanced planning and coordination. Many officials noted that participation in NEHI has frequently given regulators the opportunity to become aware of and involved with research projects at a very early point in their development, which has resulted in research that better suits the needs of regulatory agencies.

Many officials also cited the dedication of individual NEHI working group representatives, who participate in the working group in addition to their regular agency duties, as critical to the group's overall effectiveness. A number of the members have served on the body for several years, providing stability and continuity that contributes to a collegial and productive working atmosphere. In addition, because nanotechnology is relatively new with many unknowns, these officials said the agencies are excited about advancing knowledge about nanomaterials and contributing to the informational needs of both regulatory and research agencies. Furthermore, according to some officials, there is a shared sense among NEHI representatives of the need to apply lessons learned from the development of past technologies, such as genetically modified organisms, to help ensure the safe development and application of nanotechnology.

In closing, Mr. Chairman, while nanotechnology is likely to affect many aspects of our daily lives in the future as novel drug delivery systems, improved energy storage capability, and stronger, lightweight materials are developed and made available, it is essential to consider the potential risks of this technology in concert with its potential benefits. Federal funding for studying the potential EHS risks of nanotechnology is critical to enhancing our understanding of these new materials, and we must have consistent, accurate, and complete information on the amount of agency funding that is being dedicated to this effort. However, this information is not currently available because the totals reported by the NNI include research that is more focused on uses for nanotechnology, rather than the risks it may pose. Furthermore, agencies currently have limited guidance on how to report projects with more than one research focus across program component areas, when appropriate. As a result, the inventory of projects designed to address these risks is inaccurate and cannot ensure that the highest-priority research needs are met.

Mr. Chairman, this concludes my prepared statement. I would be happy to respond to any questions that you and other Members may have.

REFERENCES

[1] 1GAO, Nanotechnology: Better Guidance Is Needed to Ensure Accurate Reporting of Federal Research Focused on Environmental, Health, and Safety Risks, GAO-08-402 (Washington, D.C.: Mar. 31, 2008).

[2] 2As of December 2007, a total of four working groups exist within the NSET subcommittee: (1) Global Issues in Nanotechnology; (2) Nanotechnology Environmental and Health Implications; (3) Nanomanufacturing, Industry Liaison, and Innovation; and (4) Nanotechnology Public Engagement and Communications.

[3] 3The 21st Century Nanotechnology Research and Development Act, Pub. L. 108-153 (2003).

[4] 4GAO, Results-Oriented Government: Practices That Can Help Enhance and Sustain Collaboration among Federal Agencies, GAO-06-15 (Washington, D.C.: Oct. 21, 2005).

In: Nanotechnology in the USA: Developments, Policies and Issues ISBN 978-1-60692-800-4
Editor: Carl H. Jennings

Chapter 6

ENGINEERED NANOSCALE MATERIALS AND DERIVATIVE PRODUCTS: REGULATORY CHALLENGES*

Linda-Jo Schierow

ABSTRACT

Scientists and engineers can now examine, design, and manipulate materials at the molecular level, termed "nanoscale," between 1 and 100 billionths of a meter. The U.S. government has invested heavily to ensure that American industry remains a global leader in the field, because the products of nanotechnology are seen to have great economic potential and offer possible solutions to national problems ranging from energy efficiency to detection of agents of biological warfare. Optimism about nanotechnology is tempered, however, by concerns about the unknown potential of nanoscale materials to harm the environment and human health. Some have called for federal regulation of potential environmental, human health, and safety (EHS) risks, arguing that the lack of federal EHS regulations increases the risks of unanticipated adverse consequences due to human or environmental exposure to engineered nanomaterials. The cost of such consequences would depend on their actual, as well as publicly perceived, severity, frequency, and reversibility. The cost to the nanotechnology industry could be great, if consumers responded to a potential threat of harm by indiscriminately rejecting all products of nanotechnology, rather than the offending nanomaterial or an individual application. Others oppose federal regulatory requirements, arguing that they might unnecessarily delay the environmental, health, and economic rewards expected from nanotechnology.

Questions about the need for, and ideal form of, nanotechnology regulations are exceedingly difficult to address, given the current state of scientific understanding of engineered nanoscale materials. This chapter considers certain challenges faced by scientists, entrepreneurs, and government officials involved with nanotechnology research, as they strive to define the characteristics of nanomaterials, the potential EHS risks, and how they might be addressed. Challenges include the wide variety of nanomaterials and applications; lack of basic information about their properties; lack of

---

* This is an edited, excerpted and augmented edition of a CRS Report RL34332, dated July 18, 2008.

conventions for naming, measuring and identifying nanomaterials; the proprietary nature of some critical information; the need to prioritize federal resource needs; and a possible lack of clear statutory authority or appropriate regulatory framework to anticipate or respond to any identified risks. For more information about the national nanotechnology research agenda, appropriations, and authorizing legislation, see CRS Report RL34401, The National Nanotechnology Initiative: Overview, Reauthorization, and Appropriations Issues.

These difficulties may be surmounted over time without significant legislative action, or Congress may choose to intervene. If it does, it might choose any of several approaches. Possible approaches include increasing funding for workshops in standardization or other research relevant to identifying and possibly ameliorating any environmental or human health and safety concerns associated with nanomaterials; changing the allocation of research money among agencies or the interagency research management structure; adopting a national or international research strategy; or enacting legislation that authorizes, mandates, or constrains agency actions to require information collection or to restrict production, sale, use, or disposal of nanomaterials. Each risk management approach has potential positive and negative consequences that Congress may want to consider.

INTRODUCTION

U.S. scientists and engineers who are working at the molecular level, or nanoscale, are developing novel materials and derivative products at a rapid pace. The unique physical, chemical, and biological properties of engineered nanoscale materials lend themselves to a huge array of applications that, some analysts believe, will transform industries, foster sustainable economic growth, deliver more effective treatments for chronic diseases, and vastly improve energy efficiency. Many nations, including the United States, are eager to lead this nanotechnology revolution, and to reap its benefits. The European Union has been particularly active, but there also is intense activity in Japan, China, and other nations.

In the United States and some other nations, enthusiasm and investment in nanotechnology are somewhat restrained, however, by questions about the possible environmental, human health, and safety (EHS) risks associated with this new technology. Does nanotechnology pose risks to human health or the environment that are not being adequately controlled? If so, how will consumers here and abroad react if possible hazards are identified? Should commerce in nanomaterials or associated products be subjected to some level of government regulation? If so, do federal agencies have sufficient statutory authority, expertise, and resources to regulate potential EHS risks of engineered nanoscale materials and derivative products? The answers to such questions may determine the nature, timing, distribution, and extent of the social and economic costs and benefits associated with nanotechnology.

Some groups are calling on Congress to regulate engineered nanoscale materials and derivative products to control potential EHS risks, arguing that lack of federal regulation might increase the risks of unanticipated adverse consequences. The cost of such consequences would depend on their actual, as well as publicly perceived, severity, frequency, and reversibility. The cost to the nanotechnology industry also could be great, if consumers responded by indiscriminately rejecting all products of nanotechnology, rather than the offending nanomaterial or an individual application. Others oppose federal regulatory requirements, arguing that they might unnecessarily delay the environmental, health, and economic rewards expected from nanotechnology.

Questions about the need for, and ideal form of, regulation for nanotechnology are exceedingly difficult to address, given the current state of scientific understanding of engineered nanoscale materials. The purpose of this chapter is to consider certain challenges faced by federal EHS risk assessors, risk managers, and policy makers, and to discuss possible legislative approaches to address those challenges. For more information about the national nanotechnology research agenda, appropriations, and authorizing legislation, see CRS Report RL34401, *The National Nanotechnology Initiative: Overview, Reauthorization, and Appropriations Issues.*

The Nature of Nanotechnology. "Nanotechnology" encompasses a broad range of techniques for producing and manipulating tiny particles, thin films, and other materials at such minute dimensions that quantum effects have a measurable influence on the constituent atoms.[1] At this scale, the basic chemical, physical, and biological properties of materials can vary with slight increases and decreases in dimensions between 1 and 100 billionths of a meter. For example, slightly smaller or larger nanomaterials may be more or less magnetic or able to conduct electric currents, or they may absorb and reflect different wavelengths of light. Thus, for example, nanoparticles of gold can be red, yellow, or blue, depending on size and shape. Even when the properties of nanoscale and bulk materials are similar, they may be enhanced at the nanoscale because of the very high surface area of nanoparticles relative to their total volume. Thus, for example, relatively small doses of therapeutic drugs contained in nanoparticles may be more effective than larger doses of the same drugs contained in larger particles.

Nanotechnology in the United States. The ability to manipulate molecules to exploit particular properties promises a wealth of potential applications. The United States is a leader in the field, with many patents for commercial applications of nanotechnology granted and pending and hundreds of products incorporating nanoengineered materials being marketed. Currently available products that incorporate nanomaterials include certain cosmetics, sunscreen, tennis balls, food additives, clothes washers, and odor-free clothing. According to experts, anticipated products of nanotechnology range "from faster-burning rocket fuel additives to new cancer treatments, filters to assist in cleaning the environment, and remarkably accurate and simple-to-use detectors for biological toxins such as anthrax."[2]

To encourage and coordinate nanotechnology research and development in the United States, the President established the interagency National Nanotechnology Initiative (NNI). Launched in the President's FY2001 budget request, the NNI was codified and further defined when Congress enacted the 2 1st Century Nanotechnology Research and Development Act (Public Law 108-153) in December 2003. In accordance with the act, the President's National Science and Technology Council, through its Subcommittee on Nanoscale Science Engineering and Technology (NSET), oversees planning, coordination, and management of the National Nanotechnology Program (NNP).[3] The law requires the NNP to set goals, priorities,

Table 1. FY2007 Actual Budget for the National Nanotechnology Initiative (NNI) and Environmental, Health, and Safety (EHS) Research

(dollars in millions)

Agency/Department	NNI	EHS[a]
Department of Defense	$450	0.0
National Science Foundation	389	26.9
Department of Energy	236	0.0
Department of Health and Human Services (National Institutes of Health and National Institute for Occupational Safety and Health)	222	13.3
Department of Commerce (National Institute of Standards and Technology)	88	0.9
National Air and Space Administration	20	0.0
Environmental Protection Agency	8	7.1
Department of Agriculture (Forest Service and Cooperative State Research, Education, and Extension Service)	7	0.1
Department of Homeland Security	2	0.0
Department of Justice	2	0.0
Department of Transportation	1	0.0
Total	$1,425	$48.3

Source: National Science and Technology Council, Committee on Technology, Subcommittee on Nanoscale Science, Engineering, and Technology. National Nanotechnology Initiative: FY2009 Budget & Highlights, page 2.

a. EHS funding also is included in total NNI funding.

Scientific concern is based in part on some of the very properties that researchers hope to exploit. For example, scientists hope to use certain nanoparticles to deliver medicine to infected tissues where it can best fight disease with a minimum of unintended side effects. The small size of nanoparticles may allow them to pass easily through the skin and internal membranes. This raises questions, however, of whether exposure to nanoparticles can be effectively confined to targeted tissues, or whether environmental releases could be captured, removed from environmental media, or rendered harmless. Similarly, while high surface-area-to-mass ratio may allow nanoparticles to deliver potent doses of medicine in tiny packages, it also might amplify any toxicity of particles inadvertently encountered.[6]

It is too soon to know whether such questions are serious cause for concern, but there is scientific evidence that some nanoparticles may be hazardous. For example, certain nanoparticles are known to be toxic to microbes,[7] and EPA has reported some studies that have found nanoparticles generally (but not always) are more toxic than larger particles of identical chemical composition.[8] Other studies indicate that some nanoparticles are toxic in a way that cannot be explained by differences in particle size alone.[9] Yet, such studies are rare, and nanoparticles are diverse, so that one study with one kind of particle may not be informative with respect to the properties of other kinds of particles. Moreover, scientists have demonstrated that toxic nanoparticles may sometimes be made nontoxic by changing the surface chemistry of the particles — for example, by oxidizing the exposed atoms.[10]

The unknown potential of individual nanomaterials to harm the environment or human health might lead to consumer rejection of the entire range of consumer products incorporating nanotechnology, especially if consumers perceive that there is inadequate federal oversight. As explained by one witness who testified before the House Committee on Science, "The perception that nanotechnology will cause environmental devastation or human disease could itself turn the dream of a trillion- dollar industry into a nightmare of public backlash."[11] To prevent a loss of consumer confidence, academic researchers, policy analysts, and some entrepreneurs in nanotechnology have been working with federal agencies that have responsibility for protecting the environment, workers, and consumers. The remainder of this chapter describes some of the challenges faced by these groups as they strive to define the characteristics of nanomaterials, the risks they might pose, and how possible risks might be addressed under existing statutory authorities.

REGULATORY CHALLENGES

The Environmental Protection Agency (EPA), Food and Drug Administration (FDA), Consumer Product Safety Commission (CPSC), and Occupational Safety and Health Administration (OSHA) are actively exploring the health and safety implications of nanotechnology and the possible need for regulations. Other federal agencies also are doing research on environmental, health, and safety applications or implications of nanotechnology.[12] They face many challenges, including those discussed below.

Diversity of materials and applications. Nanomaterials vary widely in size, structure, properties, and atomic or molecular identities (that is, chemical composition). Some are relatively simple materials, composed primarily of a single element in a particular crystal form, such as carbon nanotubes. But, even carbon nanotubes may be of various lengths and thicknesses, and may be relatively pure, containing few unneeded elements, or contaminated by unknown substances. The properties being explored or exploited by researchers and developers of products may result from any combination of these features, which may vary from batch to batch supplied by carbon nanotube manufacturers or distributors.[13] For example, carbon nanotubes may be "doped" to deliberately include other substances to obtain a particular electric charge or other property. Alternatively, a core nanomaterial may be coated or covered by a nanoscale film, embedded in plastic, or otherwise modified. Some carbon nanotubes are specifically treated to prevent agglomeration into larger particles.

Many other elements and compounds may be used to produce materials through nanoengineering, and some are considerably more complex than carbon nanotubes. Currently, most commercial products fall into four categories: nanotubes (which may be carbon, silicon, or another substance); metal oxides; quantum dots; and naturally occurring clays.[14] The physical, electrical, magnetic, and other properties of these different materials vary due to chemical composition, but also due to overall dimensions and shapes of particles. Some products of nanoengineering do not even consist of nanoscale materials, but rather incorporate spaces that are nanoscale.

The risk associated with these diverse materials may depend more on the application than on the material, and the potential uses of nanomaterials are countless. For example, because risk varies with degree of exposure, risk posed by nanomaterial is likely to vary depending on whether it is embedded in plastic or some other substance that might reduce exposure, is free-standing and easily dispersed through air or water, or is coated with a more biologically active organic molecule in order to enhance exposure. Moreover, risks would be expected to vary throughout the life cycle of a product, from manufacture through use, recycling, treatment, or disposal. Thus, the potential risk from nanomaterial in cosmetics may be greater or less than the risk of the same material washed into swimming pools or lakes. In addition to the potential risks of routine manufacture, use, and disposal, risks associated with accidental, even potentially catastrophic, releases should be considered.

This diversity means that traditional regulatory toxicology and risk assessment, which typically proceed chemical by chemical, would be prohibitively time- consuming and expensive. Thus, Vicki Colvin, Executive Director of the International Council on Nanotechnology (ICON) at Rice University, proposes a different approach, which proceeds by correlating material properties with effects on the environment and human health to determine the general factors that affect toxicity. Colvin calls this risk forecasting.[15]

Lack of data characterizing nanomaterials. The structure and chemical composition of a small sample of nanomaterials produced in a laboratory can be well understood and defined, at least for relatively simple materials like buckyballs. However, the consistency of structure and chemical composition within and between batches of manufactured nanomaterials varies widely. It is possible to measure the numerous properties of nanomaterials, but it is difficult and expensive, so properties other than the ones of particular interest are not known. For example, researchers generally do not investigate a nanomaterial to determine the temperature at which it will melt or boil, or the degree to which it is soluble in water or any other solvent.

Even among researchers whose interest focuses on toxicity, there is no agreement about which data might be useful, and therefore few data are collected. Scientists have not yet determined which physical-chemical properties (for example, size, shape, composition, stability, or electric charge) will be most important in determining ecological and toxicological properties. For example, at a recent meeting of researchers interested in studying toxicity, they agreed only that it is probably most important to determine a material's surface reactivity (a rather vague notion of how readily surface molecules combine with other substances to which they are exposed, that would be measured in various ways depending on the material).[16] In addition, they generated a long list of properties of possible interest and a shorter list of properties that definitely should be investigated before toxicity is assessed.[17]

Until data are routinely collected on a basic set of physical and chemical properties, there will be no basis for hypothesizing about relationships between size, structure, chemical composition, and toxicity, or for predicting toxicity of similar, newly created substances.

Lack of standardization in nomenclature, metrics, and materials. A major obstacle to data collection is the absence of consensus on how the materials should be named, how scientific tests should be conducted, or even what constitutes a sample of a particular material. Naming conventions; standard, validated scientific methods; and standard samples of materials must be developed and made available to researchers, before the results of scientific tests will be accepted by others as valid measures and comparable across researchers and materials. If such standards were internationally accepted, it might permit international collaboration and data sharing, and speed development of an adequate data set for generalizing about nanomaterials.

The U.S. approach to standards development is voluntary. The National Institute of Standards and Technology (NIST), a non-regulatory federal agency within the U.S. Department of Commerce, is facilitating the development of a measurement system and nomenclature for use by nanotechnology scientists and engineers.[18] The NIST Center for Nanoscale Science and Technology (CNST) is

> dedicated to partnering with interested parties from industry, academia, and government to achieve common goals.... By offering collaborative opportunities, the research program also offers access to nanoscale measurement and fabrication capabilities not elsewhere available. The CNST also offers access to the CNST Nanofab, operated by professionals dedicated to serving users and offering access to state-of-the-art tools within an economical cost-sharing model.[19]

EPA, the National Institute for Occupational Safety and Health, and other U.S. agencies participating in the NNP also are working with the American National Standards Institute (ANSI); ASTM International (an international organization that uses a consensus approach to developing voluntary standards); the Nanotechnology Characterization Laboratory (established by the National Cancer Institute (NCI), NIST, and the U.S. Food and Drug Administration (FDA), to characterize nanoparticles intended for cancer therapies and diagnostics); the International Organization for Standardization (ISO); and other groups on these basic issues of nomenclature, characterization, and measurement, which must be resolved prior to toxicity data development.[20] The U.S. government is cooperating with its trading partners in the Organization for Economic Cooperation and Development to ensure development of standards that are consistent internationally.

Among the highest priorities for EHS risk assessment is development of physical standards, that is, reference materials for each nanomaterial of interest. Physical reference standards are needed to allow identification of materials being examined. Without standard samples of materials for comparison, materials being studied cannot be identified with precision, making research results impossible to interpret.[21]

Vicki Colvin, executive director of ICON, has argued that experts repeatedly identify the development of standards for conducting and reporting research as the critical first step in EHS research for nanotechnology.[22] They need protocols that specify, for example, what constitutes a toxicologically relevant dose, or whether chemical purity is a critical property, so that research reports will provide information useful to EHS risk analysts. Such standards could be developed through workshops, but there is no federal funding for such workshops, she contends.[23] Others have suggested that funding is necessary to permit travel to workshops by academics and federal employees.[24] This concern is addressed in bills to reauthorize the NNI (H.R. 5940, as passed by the House, and S. 3274, as introduced).

Proprietary nature of information. Much of the on-going research and development of nanomaterials is being conducted by private entities with an economic interest in protecting information about their work. These entities generally will not voluntarily reveal details about production processes or even the chemical composition or physical structure of their nanomaterials, due to concerns about competition, potential effect of regulatory decisions, and potential liability. Furthermore, due to the very technical and often resource-intensive nature of nanotechnology development, scientists working for private entities generally are familiar with a limited set of nanomaterials: while one laboratory studies carbon nanotubes, another might focus exclusively on metal oxides, or even on a single metal oxide. This means that scientists generally do not have access to data that are needed to detect patterns in the relationships between toxicity and other characteristics of various nanomaterials. Without such data, there is no basis for building theoretical models for hypothesis testing. In short, the proprietary nature of nanotechnology arguably impedes the scientific study of nanoscale matter, and nanotoxicology in particular, by discouraging data sharing.

Difficulty of communicating among academic disciplines. Only a few laboratories have been able to generate data for diverse categories of nanomaterials, and none has access to information about the full spectrum of materials in development. There is some hope, however, that scientific understanding of nanomaterials might be advanced by augmenting data on synthesized materials with available data on naturally occurring or incidentally produced nanomaterials, such as those found in dust or diesel exhaust. A few years ago, more than 500 peer-reviewed publications were available on naturally occurring nanoparticles. In addition, there were more than 10,000 peer-reviewed articles on incidental nanoparticles that result largely as byproducts of human activities such as mining, cooking, and metal working.[25] On the other hand, synthesized nanomaterials vary in many ways from those that are naturally occurring.[26]

Accessing data is one problem, but understanding the meaning of data across academic disciplines is another. At an EPA workshop on characterizing nanoparticles, toxicologists and physicists struggled to express their concerns to one another, and admitted frankly their ignorance of the others' areas.[27] Agreement on common terminology is likely to help, but the lack of commonality is deeper than terminology. Perhaps in time, scientists collaborating routinely at interdisciplinary research centers (and occasionally at workshops) may help to bridge the gap.

Limited resources. There are limited federal resources available to evaluate EHS implications and regulate nanomaterials, because the overall budgets of the executive agencies that are responsible for monitoring and regulating potential risks to human health and the environment have been steady or declining in recent years, while the agencies' areas of responsibility have grown.[28] For example, according to an analysis by the Congressional Research Service of data provided by the President's Office of Management and Budget, EPA's overall budget authority has remained relatively flat for the past 20 years, and has declined slightly since 2003. During the same period, Congress enacted legislation that expanded the agency's duties, and the Superfund tax authority expired. As the Superfund was depleted, EPA's budget absorbed the costs of cleaning up hazardous waste sites on the National Priority List.

Recent hearing testimony reveals an equally constrained budget situation at the Consumer Product Safety Commission (CPSC).[29] "While the CPSC has thus far been successful at facing these new and evolving challenges with diminishing resources, the 2008 funding level will challenge the Commission's ability to maintain its existing level of standards development, enforcement, public information, and international activities."[30]

The Food and Drug Administration (FDA) also faces resource constraints.[31,32] Its funding issues have been summarized in the proceedings of a workshop addressing FDA challenges generally that was convened by the Institute of Medicine.[33] Workshop participants agreed that "the Administration should request and Congress should approve substantially increased resources in both funds and personnel" for FDA.[34] Two organizations were formed in 2006 to advocate for more FDA funding across the board (that is, not just for nanotechnology).[35] A recent report by the Subcommittee on Science and Technology of FDA's Science Board concluded with respect to all FDA programs (again, not just nanotechnology) "that science at the FDA is in a precarious position: the Agency suffers from serious scientific deficiencies and is not positioned to meet current or emerging regulatory responsibilities." According to the Subcommittee, those deficiencies stem from the growth in demands on the agency without commensurate growth in resources.[36]The report states:

- The demands on the FDA have soared due to the extraordinary advance of scientific discoveries, the complexity of the new products and claims submitted to FDA for pre-market review and approval, the emergence of challenging safety problems, and the globalization of the industries that FDA regulates.
- The resources have not increased in proportion to the demands. The result is that the scientific demands on the Agency far exceed its capacity to respond. This imbalance is imposing a significant risk to the integrity of the food, drug, cosmetic and device regulatory system, and hence the safety of the public.[37]

On the other hand, within the constraints of the overall federal budget and overall budgets of the 26 National Nanotechnology Initiative (NNI) agencies, the President's Office of Management and Budget and the Congress have been encouraging agencies to allocate increasing portions of their budgetary authorities to research related to nanotechnologies.[38] A total of $1.35 12 billion was appropriated for these programs in FY2006, according to the President's Office of Management and Budget.[39] However, less than three percent of that funding was allocated to research on the potential "applications and implications of nanotechnology" for the environment and human health and safety (EHS).[40,41] The regulatory agencies are responsible for a small fraction of this EHS research funding. EPA's EHS budget for nanotechnology in FY2006 was $3.7 million, and some of this funding was directed toward development of environmentally beneficial applications of nanotechnology, for example, to remove arsenic from surface water, rather than research relevant to evaluating potential toxicity. EPA was the only regulatory agency identified as a contributor to NNI funding in the President's budget.[42] However, it should be noted that given the rudimentary understanding of nanomaterials, it is not surprising that a relatively large portion of research funding goes to basic scientific studies. Such research is funded by the National Science Foundation, which receives the bulk of the EHS research budget, some $21 million in FY2006. In addition, substantial research is conducted to develop standardized tools and measures of nanomaterials and to understand their interactions with living things.

Possibly inadequate statutory authority. A final potential obstacle to federal risk management for nanotechnology is a lack of clear statutory directives or appropriate regulatory frameworks to guide federal risk managers. Although the Bush Administration[43] and several legal reviews of existing environmental, health, and safety statutes have concluded that they probably provide adequate authority for federal regulators over nanotechnology, such laws were not written with nanomaterials in mind.[44] As a result, agencies would have to develop new policies, produce guidance, and possibly issue regulations to translate statutory requirements with respect to nanomaterials. These tasks almost certainly would be controversial, because agencies would be making decisions that might, on the one hand, delay or restrict commerce or, on the other hand, allow entrepreneurs to market products whose effects on health or the environment are unknown or uncertain.

One concern about existing environmental, health, and safety statutes is that most apply to a specific category of chemical products intended for a particular application, for example, as a food additive, drug, cosmetic, pesticide, or consumer product. This might lead to redundant or inconsistent regulation of a nanomaterial under more than one federal law. For example, under the Federal Insecticide, Fungicide, and Rodenticide Act (FIFRA),[45] EPA must evaluate and regulate commerce in, and use of, all products that are intended to control pests, including bacteria. EPA already has stated its intent to regulate nanosilver under FIFRA when it is released from certain washing machines and other products for which manufacturers claim antibacterial properties. Other EPA program offices (the Offices of Air and Radiation, and of Water, for example) also have responsibilities for managing nanomaterials, including nanosilver, under certain conditions. To address this challenge, EPA's Science Policy Council, an internal policy group, formed a Nanotechnology Workgroup in December 2004 and charged it with describing "key science issues EPA should consider to ensure that society accrues the important benefits to environmental protection that nanotechnology may offer, as well as to better understand any potential risks from exposure to nanomaterials in the environment."[46] The Toxic Substances Control Act (TSCA),[47] which applies to all categories of chemical uses not otherwise regulated, also allows coordination to reduce any potential regulatory burden. TSCA Section 9(d) requires that the EPA

Administrator achieve "the maximum enforcement of [TSCA] while imposing the least burdens of duplicative requirements on those subject to the Act."

On the other hand, current laws sometimes exclude certain nanomaterials from requirements. For instance, TSCA Section 8(b)(1) clearly excludes from its requirements substances that are produced and used only in research laboratories. This exclusion might apply to most of the various nanomaterials currently in existence. Less clearly, TSCA excludes nanomaterials that are not "chemical substances" as defined in the law.[48] TSCA Section 2 defines a "chemical substance" as "any organic or inorganic substance of a particular molecular identity" that is not a mixture. Based on this definition, it might not be clear whether certain nanoparticles consisting of a core inorganic material coated by an organic material would qualify as a TSCA "chemical substance" or a mixture. Other nanomaterials, like nanotubes or fullerenes, have clear chemical identities in terms of chemical composition and crystal structure, but have variable properties due to differences in size or shape of particular particles. Size and shape are not normally considered in identifying molecular identity, or in distinguishing one chemical substance from another. Therefore, EPA has indicated that it does not consider size a relevant feature under TSCA.[49,50] But, of course, size is a central issue with nanomaterials.

In some cases, it is the regulations rather than the statute itself that complicate agency decisions and actions with respect to nanomaterials. For example, laws often direct agencies to exclude from regulatory requirements small quantities of chemicals, particularly chemicals not yet in commerce. If the agencies define "small quantities" in terms of weight, as EPA does for purposes of the periodic TSCA inventory updates, nanomaterials may well be excluded, because few are produced in large quantities by weight. Another example might be the exclusion of nanomaterials from food additive regulations, if the Food and Drug Administration were to decide that they fit into a category normally exempted, such as that for substances "Generally Recognized As Safe" (GRAS). Under sections 201(s) and 409 of the Federal Food, Drug, and Cosmetic Act, this includes any substance that is "generally recognized, among qualified experts, as having been adequately shown to be safe under the conditions of its intended use."[51] For more information on the possible limitations of existing laws, see reports issued by the Environmental Law Institute, the American Bar Association, and the Woodrow Wilson International Center for Scholars' Project on Emerging Nanotechnologies.[52]

VOLUNTARY INITIATIVES

As agencies consider the possible need for and shape of regulations, some stakeholders are voluntarily engaging in what is called "responsible development" of nanotechnologies. Professional organizations, industries, universities, environmental organizations, and government have been involved in such efforts. A few of the better known initiatives are described below.

The IEEE (formerly the Institute of Electrical and Electronics Engineers) is developing standard methods needed to mass produce and market electronics and photonics products while protecting workers and addressing environmental concerns.[53]

Intel, DuPont, and other large companies voluntarily adhere to responsible principles that have served in the past to minimize EHS problems associated with production of materials that are not nanoscale. They complain, however, that the adequacy of such practices with respect to nanotechnologies is unknown, and urge the federal government to sponsor additional research to "shed more light on what the best approach to protecting health and safety should be." The National Institute for Occupational Safety and Health (NIOSH) is working with industry to gather data on exposure and worker health that should help guide the design of studies in occupational settings.

In its 2007 White Paper, EPA expressed the view that

> partnerships with industrial sectors will ensure that responsible development is part of initial decision making. Working in partnership with producers, their suppliers, and users of nanomaterials to develop best practices and standards in the workplace, throughout the supply chain, as well as other environmental programs, would help ensure the responsible development of the production, use, and end of life management of nanomaterials.[54]

In that spirit, EPA worked with stakeholder groups to develop a voluntary Nanoscale Materials Stewardship Program (NMSP). The program, which pertains to engineered nanoscale materials that are in commerce or about to enter commerce, was launched January 28, 2008.[55]

The NMSP allows two levels of participation, basic and in-depth. Under the basic program, EPA will collect available data and information from manufacturers and processors of existing chemical nanoscale materials. In addition, EPA will ask participants to identify their risk management practices and to develop a risk management plan.[56] Participants in the in-depth program will develop new test data needed to provide a firm scientific foundation for future work and regulatory/policy decisions. The agency intends to use the information gained from the stewardship program to guide development of its TSCA program for nanoscale materials.

Another voluntary initiative produced a guidebook for responsible corporate behavior called the Nano Risk Framework. The six-point program was developed by DuPont Corporation working in partnership with Environmental Defense, an advocacy group, and was announced June 21, 2007, at a seminar sponsored by the Woodrow Wilson International Center for Scholars, Project on Emerging Nanotechnologies. The framework presents a process "for identifying, managing, and reducing potential environmental, health, and safety risks of engineered nanomaterials across all stages of a product's 'lifecycle.'"[57]

A final example of a voluntary initiative that aims to promote responsible development of nanotechnology is spearheaded by the International Council on Nanotechnology (ICON) at Rice University. As described on its website, "ICON is a technically driven organization whose activities are broadly supported by industry, non-profit foundations, and governments. Its multi-stakeholder partnerships and governance, with members that span the globe, make it uniquely positioned to ensure global coordination and cooperation in nanotechnology risk management."[58] Its mission is "to develop and communicate information regarding potential environmental and health risks of nanotechnology, thereby fostering risk reduction while maximizing societal benefit."[59] ICON encourages close work between developers of nanotechnology and toxicologists. As explained by Vicki Colvin, executive director of ICON, "If we understand why a material is cytotoxic [that is, toxic to cells], we should be able to make it less reactive and knock out its toxicity...."[60] Thus, she advises, "chemists making systematic changes in materials must work with people who can measure their biological effects. Tight collaboration between materials engineers, chemists, and toxicologists could provide the essential data that can enable us to engineer safer nanomaterials from the beginning."[61]

LEGISLATIVE OPTIONS

The regulatory challenges posed by nanomaterials and nanotechnologies may be resolved over time and to some extent without significant legislative action, as stakeholders work together, scientists learn more about processes and properties on the nanoscale, and federal regulators gradually adapt rules to implement existing statutory authorities. Congress might, therefore, continue to take a wait-and-see approach to nanotechnologies, perhaps combined with congressional oversight of agencies' activities. However, should Congress choose to intervene, a range of legislative strategies is available, as described below.

Increase and/or reallocate funding for health and safety research. The overall contribution from agencies' budgets to the National Nanotechnology Initiative (NNI) has grown substantially over the years.[62] It is more difficult to characterize trends in the allocation of funds devoted to research on the potential EHS implications and applications of nanotechnologies, although it too appears to have grown significantly.[63] In some cases, however, this growth arguably has been at the expense of agencies' other programs. This is particularly likely for the regulatory agencies, EPA, FDA, and the CPSC.[64] For example, the National Research Council observed that although there had been "pockets of increased funding for EHS-related research," including a proposed $4 million increase in the FY2007 budget for nanotechnology research within EPA, "there was a 4 percent cut in EPA's overall FY2007 budget."[65]

The National Research Council has recommended an increase in funding for research relevant to evaluating the potential health and safety risks associated with nanotechnologies, including work to develop requisite definitions, protocols, and methodologies.[66] Many companies, public interest groups, and the NanoBusiness Alliance (a trade group) also have asked Congress for additional funding for EHSrelated research.[67] According to the Chairman of the House Subcommittee on Research and Science Education of the Committee on Science, "The basic position of most outside observers from industry and non-governmental organizations is that the funding level should be on the order of 10% of the initiative's total funding, rather than the current 4%."[68] H.R. 5940, as passed by the House, and S. 3274, as introduced, would reauthorize the National Nanotechnology Initiative and require an official in the Office of Science and Technology to oversee planning and budget requests for EHS research, but would not require that participating agencies allocate a set percentage of nanotechnology funding for EHS research.

Congress also might wish to consider whether to change the allocation of research money among agencies. For example, it might wish to increase or decrease basic research through the National Science Foundation relative to research that might inform risk assessments at the National Institutes of Health. See Table 1, above, for the distribution of the FY2007 actual budget among agencies and departments of the NNI.

It is difficult to assess the need for additional federal funding or the adequacy of its allocation among agencies without detailed information about research priorities. Many policy analysts have argued for several years that the foremost need with respect to nanotechnology-related EHS research is a strategy or plan "to avoid duplication of research and to set priorities."[69] The House Committee on Science asked repeatedly for the NNI to develop such a strategy. In September 2006, the NNI delivered to the Committee a general framework for EHS research, which was developed by the interagency Nanotechnology Environmental and Health Implications (NEHI) Working Group.[70] The report identified five research categories and some specific needs within each. The five research categories include

- instrumentation, metrology, and analytical methods;
- nanomaterials and human health;
- nanomaterials and the environment;
- health and environmental surveillance; and
- risk management methods.

Some experts who have advocated for a stronger federal role in managing the risks of nanotechnology argued that this NNI categorization provided insufficient direction for managers, researchers, and research grant authorities. They preferred a more "top- down" approach to EHS research management, in order to ensure that the information being collected is most useful for risk managers.[71] Their priorities were published in November 2006.[72] Some of the same individuals co-authored a paper published in November 2007 that was based on a workshop held in April 2006.[73] That paper identified six critical information needs for evaluating and predicting the toxicity of nanoparticles:

- extensive physico-chemical characterization;
- capacity for macromolecular perturbation (for example, for interfering with repair of DNA or with proteins important to the immune system);
- potential for unintended carriage of toxic molecules;
- translocation (for example, from the surface of skin into the blood stream);
- agglomeration state; and
- chemical composition.

The Director of the National Nanotechnology Coordination Office and the Co- Chair of the President's Council of Advisors on Science and Technology (PCAST) disagree that a top-down approach is needed, arguing that no single individual could have the breadth of expertise necessary to adequately oversee all aspects of nanotechnology EHS research.[74] Rather, an interagency working group can "cast the wide net necessary to address the array of nanotechnology-related EHS issues," and this "interagency process will lead to a sound research strategy."[75]

Proponents of the top-down approach have found these arguments unconvincing, arguing instead that a top-down approach can involve many agencies and other stakeholders. Environmental Defense, a group that advocates for responsible development of nanotechnology, has suggested that NNI responsibilities are potentially in conflict, because they include EHS oversight of research and development on the one hand, and promotion of nanotechnology research and development on the other. Thus, the group argues, some portion of the NNI should be given "independent budgetary and management authority, responsibility, accountability, and sufficient resources to develop and direct the overall Federal nanomaterial risk research strategy."[76] Environmental Defense modeled this proposal on the approach taken by the federal government with respect to nuclear power.[77]

The Senate Committee on Appropriations expressed its desire for a research strategy in S.Rept. 110-91, which accompanied S. 1696, a bill providing FY2008 appropriations for the Department of the Interior, environment, and related agencies.[78]

The Committee is committed to ensuring that all Federal environmental, health and safety research is prioritized and coordinated so that nanotechnology' s potential benefits to the economy and environment are realized at the same time that human health and the environment are protected. To further these goals, the Committee urges EPA to contract or enter into a cooperative agreement with the National Academy of Sciences' Board on Environmental Studies and Toxicology within 90 days of enactment to develop and monitor implementation of a comprehensive, prioritized research roadmap for all Federal agencies on environmental, health and safety issues for nanotechnology.[79]

The Senate did not act on S. 1696, but in accord with the explanatory statement for the Consolidated Appropriations Act, 2008, which became Public Law 110-161, the report language for S. 1696 is being treated as approved.[80] According to Celia Merzbacher, who formerly was co-chair of the NSTC Subcommittee on Nanoscale Science, Engineering, and Technology, therefore, the National Academy review of the NNI strategy will take place.[81] The NAS will review the final version of the NNI strategy.

The National Science and Technology Council (NSTC) released the final NNI Strategy for Nanotechnology-Related Environmental, Health, and Safety Research in mid-February 2008.[82] The strategy builds on the five research categories described above, but also identifies and prioritizes specific research needs within each category. Research under way in FY2006 was matched to the five research categories, and timelines were developed to guide future activities. Timelines reflect the agencies' immediate needs as well as their views of research capacities and prerequisites. Research in the area of instrumentation, metrology, and analytical methods is considered cross-cutting and of highest priority. This research will be coordinated by the National Institute for Standards and Technology. The National Institutes of Health will coordinate research related to human health, and the EPA will coordinate research related to the environment. NIOSH is assigned responsibility for coordinating research related to human and environmental exposure assessment. Research related to risk management will be coordinated by the Food and Drug Administration (FDA) and EPA. Finally, NSET noted that the research strategy is expected to be reviewed and updated as research progresses and needs and priorities evolve.[83]

H.R. 5940, as passed by the House, and S. 3274, as introduced, would require the Director of the President's Office of Science and Technology Policy (OSTP) to designate an associate director of OSTP as Coordinator for Societal Dimensions of Nanotechnology. The Coordinator would be responsible for 1) ensuring that a research plan for environmental, health, and safety research activities is developed, updated, and implemented; 2) "encouraging and monitoring" agencies participating in the NNI "to allocate the level of resources and management attention necessary to ensure that the ethical, legal, environmental, and other appropriate societal concerns related to nanotechnology, including human health concerns, are addressed;" and 3) encouraging agencies to identify, assess, and implement suitable mechanisms for establishing public-private partnerships to support EHS research. The bill would require the plan to 1) specify near-term research objectives and long-term research objectives; 2) specify milestones, and the time and resources needed for achieving near-term objectives; 3) describe the roles of the agencies in achieving the objectives; 4) specify the funding allocated to each objective and the sources of funding by agency; and 5) estimate the funding required and the source of funding by agency for each major objective for three future years.

Mandate/constrain reporting by manufacturers of nanotechnology. Congress also might intervene to ensure an appropriate level of information collection by regulatory agencies. If Congress wants to ensure that information about the potential risks of nanotechnology and nanomaterials is collected and does not want to rely on voluntary programs, or conversely, if Congress wants to prevent agencies from imposing reporting requirements, legislation might be necessary. Congress could direct or constrain agency action that would require manufacturers of nanotechnology materials or products to determine physical and chemical properties, to conduct toxicity tests, or to report information that already is reasonably available and potentially relevant to EHS. Either requirements or constraints could be phased into effect in order to ensure that requirements would be commensurate with risk and investments by the regulated community. For example, increasing demands for information could be tied to the introduction or marketing of new applications, new products, or threshold quantities of products. Alternatively, agencies might be instructed to refrain from requiring long-term and costly studies, at least until a certain production threshold is attained.

For some chemical substances and applications, regulatory agencies already have certain regulatory authorities. For example, under the Toxic Substances Control Act (TSCA), Section 8(d), EPA requires that manufacturers submit lists of unpublished health and safety studies known to have been conducted, and copies of such studies, on request. If that authority is not sufficient, too broad, or not clear with respect to nanomaterials, TSCA and other statutes could be amended. Alternatively, reporting and testing requirements, limitations, or prohibitions could be included in free-standing legislation. A third option might be to tie requirements for reporting or testing to legislation authorizing research funding.

These options might have unintended consequences. For example, depending on the specific provisions, new reporting or testing requirements might be considered an impediment to innovation by small or medium-sized enterprises, or too burdensome for manufacturers who embed nanomaterials in hard plastics or other substances. A variety of methods are available to reduce unintended consequences. For example, to reduce the burden imposed by a testing requirement, Congress might allow manufacturers to share test data (and costs of testing), grant exclusive production or marketing rights within the United States for a number of years to manufacturers who conduct testing (to compensate them for their expenditures), or exempt particular categories of products or manufacturers from requirements. Again, existing law provides an example of how some requirements might be tailored. The Federal Insecticide, Fungicide, and Rodenticide Act (FIFRA) directs EPA to promulgate testing and reporting requirements for pesticides, mechanisms for simplifying requirements for relatively safe pesticides, and compensation rules, and grants those who conduct tests a period of exclusive data use.

On the other hand, attempts to tailor requirements, limitations, or prohibitions might be viewed as unfair by some regulated entities, if they are perceived to treat manufacturers differently, conferring advantages on some but not others. Moreover, any exemptions for small or medium-sized enterprises would reduce the amount of information collected, information that might be important to risk assessors.

A potential benefit of requiring testing or reporting is that useful information might become available to the regulatory agencies, allowing them to better evaluate the significance of potential EHS risks or to assess the potential value of benefits. This could lead to regulations that were more focused, reasonable, and economically efficient, because agencies could target regulations toward technologies or products posing greater relative risks (and perhaps smaller benefits).

Clarify, enlarge, or restrict agencies' authority to regulate. Congress also might legislate to ensure that nanotechnology would, or would not, be regulated to manage any EHS risks that might be identified. Congress could either authorize or restrict agencies' authorities to regulate any stage in the lifecycle of nanomaterials: production, sale, use, or disposal.

Imposing requirements on manufacturers might delay the environmental, health, and economic rewards expected from nanotechnology. At the same time, EHS regulations might reduce any risk of adverse consequences from exposure to nanomaterials. The Bush

Administration has issued guidelines for any regulations that might be imposed under existing statutes.[84]

If, on the other hand, Congress chose to prohibit or restrict agencies' authority to regulate nanotechnology or products, the potential economic benefits of the new technology might be more quickly realized, but the risks of unanticipated adverse consequences might be greater. The cost of such consequences would depend on their actual, as well as publically perceived, severity, frequency, and reversibility. The cost to companies developing nanotechnology products also could be great, if consumers responded by indiscriminately rejecting all products of nanotechnology, rather than a single offending nanomaterial or application.

Conclusion

The need for additional research to identify the potential hazards that might be associated with nanotechnology and to evaluate risks related to the environment and human health and safety (EHS) is not in dispute. However, there is a range of views about whether there is a need for increased federal intervention at this time. The regulatory challenges posed by nanomaterials and nanotechnologies are many — the diversity of nanomaterials, lack of data characterizing the materials, lack of standardization in nomenclature and metrics, the proprietary nature of private research results, limited resources in regulatory agencies, and possibly inadequate statutory authority. These difficulties may be surmounted over time without legislative action, or Congress may choose to intervene. If it does, it might choose any of several approaches. Selected approaches include increasing funding for workshops in standardization and other EHS research, changing the allocation of research money among agencies, adopting and implementing a national and/or international research strategy, or enacting legislation that authorizes, mandates, or constrains agency actions to require information collection or to restrict production, sale, use, or disposal of nanomaterials. H.R. 4040, as passed by the House, and S. 3274, as introduced, would require appointment of an official who would be required to oversee development and implementation of an EHS research plan.

It is noteworthy that Congress is considering its options at this early stage of technology development, when only a few nanomaterials are being manufactured on a large scale. Risk management decisions nonetheless are pressing, as the rate of nanotechnology development and commercialization is rapidly escalating.

References

[1] Quantum effects are the result of quantum physics, rules that predict the behavior of matter and energy at the atomic level, where the more familiar laws of classical physics do not apply.

[2] Ratner, Mark, and Daniel Ratner. 2003. Nanotechnology: A Gentle Introduction to the Next Big Idea. Prentice Hall: Upper Saddle River, NJ. p. 3.

[3] The National Science and Technology Council (NSTC) is chaired by the President and includes the Vice President, the Director of the Office of Science and Technology

Policy, Cabinet Secretaries and Agency Heads with significant science and technology responsibilities, and other White House officials. According to its website, "It is the principal means within the executive branch to coordinate science and technology policy across the diverse entities that make up the Federal research and development enterprise."

[4] National Nanotechnology Initiative. About the NNI. [http://www.nano.gov/html/about/home_about.html], visited January 10, 2008.

[5] These agencies do not distinguish work conducted on nanotechnology from other work, and do not report budget figures to the NNI. Moreover, OSHA and FDA do not conduct toxicological research, although they do apply the results of such research in risk assessments as a basis for regulatory decisions.EPA Science Advisory Board, Board of Scientific Counselors. Meeting summary, October 19-20, 2006. p. 26. National Science and Technology Council, Committee on Technology, Subcommittee on Nanoscale Science, Engineering, and Technology. The National Nanotechnology Initiative: Research and Development Leading to a Revolution in Technology and Industry. Supplement to the President's 2008 Budget, July 2007, pages 7 and 11. Hereafter cited as the FY2008 Budget Supplement.

[6] Science Policy Council, Nanotechnology Workgroup. 2007. U.S. Environmental Protection Agency Nanotechnology White Paper. EPA 100/B-07/001. U.S. EPA: Washington DC, p. 14. Hereafter cited as EPA White Paper.

[7] Silver, for example, is toxic to microbes, and some product manufacturers have made antibacterial claims for their products containing nanosilver. In addition, research has demonstrated the toxicity of C60 fullerenes to bacteria in water under laboratory conditions (Fortner, J.D., D.Y. Lyon, C.M. Sayes, et al. "C60 in water: Nanocrystal formation and microbial response," Environmental Science & Technology, v. 39, (2005), p. 4307-4316.)

[8] EPA White Paper, p. 54.

[9] Ibid.

[10] National Research Council. 2006. A Matter of Size. National Academies Press, Washington, DC. p. 157. Hereafter cited as National Research Council 2006.

[11] Colvin, Vicki L. Testimony before the Committee on Science, U.S. House of Representatives. Hearing on "The Societal Implications of Nanotechnology." 108th Cong., 1st Sess., April 9, 2003. U.S. Govt. Print. Off., Washington, DC.

[12] EPA Science Advisory Board, Board of Scientific Counselors. Meeting summary, October 19-20, 2006. p. 26.

[13] Busnaina, Ahmed. Director, National Science Foundation Nanoscale Science and Engineering Center for High-rate Nanomanufacturing, Northeastern University. Comments made at the EPA Peer Consultation on Materials Characterization of Nanoscale Materials, September 6-7. 2007, at Arlington, Virginia.

[14] Goldman, Lynn, and Christine Coussens, eds. 2005. Implications of Nanotechnology for Environmental Health Research, Institute of Medicine, National Academies Press, Washington, DC. pp. 6-7. These categories are not all inclusive, and there are other categorization schemes, but this scheme accounts for the vast majority of commercial applications in 2007.

[15] Computational toxicology presumably is a form of risk forecasting. For more on computational toxicology, see CRS Report RL341 18, The Toxic Substances Control Act (TSCA): Implementation and New Challenges, by Linda-Jo Schierow.

[16] EPA Peer Consultation on Materials Characterization of Nanoscale Materials, September 6-7. 2007, at Arlington, Virginia.

[17] 17 EPA Peer Consultation on Materials Characterization of Nanoscale Materials, September 6-7. 2007, at Arlington, Virginia. The shorter list was originally suggested by David Warheit of DuPont Haskell Laboratory and, included particle size and size distribution (wet state) and surface area (dry state) in the relevant media being utilized — depending upon the route of exposure; crystal structure/crystallinity; aggregation status in the relevant media; composition/surface coatings; surface reactivity; method of nanomaterial synthesis and/or preparation including post-synthetic modifications (e.g., neutralization of ultrafine titanium dioxide particle- types); and purity of sample.

[18] Center for Nanoscale Science and Technology, National Institute of Standards and Technology. About NIST's Center for Nanoscale Science and Technology. [http://cnst.nist. gov/about_cnst.html], visited January 9, 2008.

[19] National Institute of Standards and Technology website. General Information. [http://www.nist.gov/public_affairs/general2.htm], visited January 9, 2008.

[20] EPA Peer Consultation on Materials Characterization of Nanoscale Materials. September 6-7, 2007, Arlington, Virginia. EPA White paper, p. 32.

[21] Small, John. Group Leader, Center for Nanoscale Science and Technology, NIST. EPA Peer Consultation on Materials Characterization of Nanoscale Materials. September 6-7, 2007, Arlington, Virginia.

[22] U.S. Congress. House. Committee on Science and Technology. Subcommittee on Research and Science Education. Testimony on "Research on Environmental and Safety Impacts of Nanotechnology: Current Status of Planning and Implementation under the National Nanotechnology Initiative." Hearing, 1 10th Cong., 1st Sess. October 31, 2007.

[23] Ibid.

[24] EPA Peer Consultation on Materials Characterization of Nanoscale Materials. September 6-7, 2007, Arlington, Virginia.

[25] Goldman and Coussens, 2005, p. 7, citing Eva Oberdorster, unpublished.

[26] Naturally occurring particles will vary in shape, size, and properties, while synthesized nanomaterials are designed to be more uniform. Thus, synthesized materials often may be redesigned and re-engineered to eliminate or ameliorate problems (such as toxicity) that emerge.

[27] EPA Peer Consultation on Materials Characterization of Nanoscale Materials. September 6-7, 2007, Arlington, Virginia.

[28] However, the Administration has stated that the budget is adequate to the task.

[29] U.S. Consumer Product Safety Commission, 2008 Performance Budget Request, submitted to Congress, February 2007, page vii. [http://www.cpsc.gov/CPSCPUB /PUBS/ REPORTS/2008plan.pdf], visited January 16, 2008.

[30] Ibid.

[31] Michael R. Taylor. 2006. Regulating the Products of Nanotechnology: Does FDA Have the Tools It Needs? Project on Emerging Nanotechnologies, Woodrow Wilson International Center for Scholars, Washington, DC. pp. 46-47.

[32] For a CRS analysis of the budget situation at FDA, see CRS Report RL34334, The Food and Drug Administration: Budget and Statutory History, FY1980-FY2007 by Judith A. Johnson, Donna V. Porter, Susan Thaul, and Erin D. Williams.

[33] Institute of Medicine. 2007. Challenges for the FDA: The Future of Drug Safety, Workshop Summary. National Academies Press, Washington , DC, p. 2. [http://books.nap.edu/catalog/11969.html], visited January 16, 2008.

[34] Ibid. p. 13.

[35] Each group has a website: the FDA Alliance site is at [http://www.StrengthenFDA.org], while the Coalition for a Stronger FDA website is at [http://www.FDACoalition.org]. Both sites were visited December 14, 2007.

[36] FDA Science Board, Subcommittee on Science and Technology. November 2007. FDA Science and Mission at Risk. Report. The report is marked "confidential" but is posted on the website of the Science Board. [http://www.fda.gov/ohrms/dockets/ac/07/briefing/2007-4329b_02_01_FDA%20Report%20on%20Science%20and%20Technology.pdf], visited January 16, 2008.

[37] Ibid., p. 2.

[38] The NNI is coordinated through the White House National Science and Technology Council's Nanoscale Science, Engineering and Technology (NSET) subcommittee. NSET does not have budget authority or appropriations. Rather, each agency allocates part of its budget to nanotechnology and reports its efforts to the NNI.

[39] FY2008 Budget Supplement.

[40] Ibid.

[41] For FY2007, estimated NNI and EHS funding levels are somewhat higher, and higher still in the President's FY2008 budget request.

[42] The National Institute for Occupational Safety and Health informs regulatory activities by the Occupational Safety and Health Administration (OSHA), but is not itself an agency that issues EHS regulations.

[43] Marburger, John H. III, and James L. Connaughton. Principles for Nanotechnology Environmental, Health, and Safety Oversight, memorandum for the heads of Executive departments and agencies, November 8, 2007. [http://www.ostp.gov/galleries/default-file/ Nano%20EHS%20Principles%20Memo_OSTP-CEQ_FINAL.pdf], visited January 22, 2008.

[44] Breggin, Linda K. 2005. Securing the Promise of Nanotechnology: Is U.S. Environmental Law Up To the Job? Environmental Law Institute, Washington, DC. Hereafter cited as Breggin 2005. [http://www.elistore.org/], visited October 11, 2007.

[45] American Bar Association (ABA) Section of Environment, Energy, and Resources, Briefing documents. [http://www.abanet.org/environ/nanotech/], visited October 11, 2007.

[46] 7 U.S.C. 136-136y

[47] Science Policy Council, Nanotechnology Workgroup. 2007. U.S. Environmental Protection Agency Nanotechnology White Paper. EPA 100/B-07/001. U.S. EPA: Washington, DC. 120 pp.15 U.S.C. 2601 et seq.

[48] TSCA Sections 4 and 8 also authorize reporting requirements. For more information about TSCA, see CRS Report RL341 18, The Toxic Substances Control Act (TSCA): Implementation and New Challenges, by Linda-Jo Schierow.

[49] EPA. TSCA Inventory Status of Nanoscale Substances — General Approach, p. 4. Distributed at the EPA Peer Consultation on Materials Characterization of Nano scale Materials. September 6-7, 2007, Arlington, Virginia. Also available in the EPA docket EPA-HQ-OPPT-2004-0 122-0057 at [http://www.regulations.gov/], visited October 19, 2007.

[50] However, according to J. Clarence Davies (2007, EPA and Nanotechnology: Oversight for the 21st Century, Project on Emerging Nanotechnologies, Woodrow Wilson International Center for Scholars, Washington, DC, p. 3 1), EPA did determine that carbon nanotubes have properties different from other forms of carbon and did decide that it could be regulated under TSCA.

[51] U.S. Food and Drug Administration. Center for Food Safety and Applied Nutrition. Guidance for Industry: Frequently Asked Questions About GRAS. December 2004. [http://www.cfsan.fda.gov/~dms/grasguid.html#Q1], visited January 16, 2008.

[52] Breggin 2005.J. Clarence Davies, 2007, EPA and Nanotechnology: Oversight for the 21st Century; Breggin, Linda K., and John Pendergrass, Where Does the Nano Go? End-of-Life Regulation of Nanotechnologies; and Michael R. Taylor, 2006, Regulating the Products of Nanotechnology: Does FDA Have the Tools It Needs? Project on Emerging Nanotechnologies, Woodrow Wilson International Center for Scholars, Washington, DC. [http://www.nanotechproj ect.org/], visited January 16, 2008.

[53] National Research Council 2006, p. 153.

[54] EPA White paper, p. 63.

[55] 73 Federal Register 4861-4866.

[56] EPA. Nanoscale Materials Stewardship Program. [http://www.epa.gov/oppt/nano/stewardship.htm], visited February 14, 2008.

[57] Nanorisk Framework. [http://nanoriskframework.com/page.cfm?tagID= 1095], visited January 16, 2008.

[58] ICON website. Background. [http://icon.rice.edu/about.cfm?doc_id=4380], visited January 16, 2008.

[59] ICON website. Mission and Strategy. [http://icon.rice.edu/about.cfm?doc_id=4379], visited January 16, 2008.

[60] 60 National Research Council. 2006. A Matter of Size. National Academies Press, Washington, DC. p. 157.

[61] Ibid.

[62] FY2008 Budget Supplement.

[63] Ibid.National Research Council. 2006. A Matter of Size. National Academies Press, Washington, DC. p. 151. Hereafter cited as National Research Council, 2006.

[64] Although FDA does not conduct toxicological research, it does interpret research conducted by manufacturers and apply the results of such research in risk assessments as a basis for regulatory decisions.

[65] National Research Council, 2006, p. 91.

[66] Ibid., p. 6 - 8, 11-12, 38, 92.

[67] Ibid., p. 91.

[68] U.S. Congress. House. Committee on Science and Technology. Subcommittee on Research and Science Education. Testimony on "Research on Environmental and Safety Impacts of Nanotechnology: Current Status of Planning and Implementation

under the National Nanotechnology Initiative." Hearing, 1 10th Cong., 1st Sess. October 31, 2007. Opening statement of Chairman Brian Baird.

[69] For example, Carol Henry of the American Chemistry Council: National Research Council. 2006. A Matter of Size. National Academies Press, Washington, DC, p. 161.

[70] National Nanotechnology Initiative. 2006. Environmental, Health, and Safety Research Needs for Engineered Nanoscale Materials. Nanoscale Science, Engineering, and Technology Subcommittee, Committee on Technology, National Science and Technology Council, Washington DC. 62 p

[71] U.S. Congress. House. Committee on Science and Technology. Subcommittee on Research and Science Education. Research on Environmental and Safety Impacts of Nanotechnology: Current Status of Planning and Implementation under the National Nanotechnology Initiative. Hearing, 1 10th Cong., 1st Sess. October 31, 2007. Testimony of Andrew Maynard.

[72] Maynard, Andrew D., Robert J. Aitken, Tilman Butz, Vicki Colvin, Ken Donaldson, Günter Oberdörster, Martin A. Philbert, John Ryan, Anthony Seaton, Vicki Stone, Sally S. Tinkle, Lang Tran, Nigel J. Walker, and David B. Warheit. 2006. Safe Handling of Nanotechnology. Nature, v. 444, (16 November) pp. 267-269.

[73] Balbus, John M., Andrew D. Maynard, Vicki L. Colvin, Vincent Castranova, George P. Daston, Richard A. Denison, Kevin L. Dreher, Peter L. Goering, Alan M. Goldberg, Kristen M. Kulinowski, Nancy A. Monteiro-Riviere, Günter Oberdörster, Gilbert S. Omenn, Kent E. Pinkerton, Kenneth S. Ramos, Kathleen M. Rest, Jennifer B. Sass, Ellen K. Silbergeld, and Brian A. Wong. 2007. Meeting Report: Hazard Assessment for Nanoparticles — Report from an Interdisciplinary Workshop. Environmental Health Perspectives, v.115, n.11, (November) p. 1654-1659.

[74] U.S. Congress. House. Committee on Science and Technology. Subcommittee on Research and Science Education. Research on Environmental and Safety Impacts of Nanotechnology: Current Status of Planning and Implementation under the National Nanotechnology Initiative. Hearing, 1 10th Cong., 1st Sess. October 31, 2007. Testimony of Clayton Teague and Floyd Kvamme.

[75] Ibid. Testimony of Floyd Kvamme.

[76] Environmental Defense. Potential model for restructuring National Nanotechnology Initiative offered to better address nano risks. Press release. November 19, 2007. [http://www.environmentaldefense.org/pres srelease.cfm?contentID=7346], visited January 16, 2008.

[77] According to the Nuclear Regulatory Commission, "The NRC was created as an independent agency by the Energy Reorganization Act, signed into law October 11, 1974, which abolished the Atomic Energy Commission. The NRC, which took over the regulatory functions of the AEC, formally came into being on January 19, 1975. The Energy Research and Development Administration, also created by the Energy Reorganization Act, took over the other functions of the AEC and is now part of the Department of Energy."

[78] U.S. Congress. Senate. Committee on Appropriations. Department of the Interior, Environment, and Related Agencies Appropriations Bill, 2008. Report to Accompany S. 1696. S.Rept. No. 110-91, 1 10th Cong., 1st Sess. Washington, U.S. Govt. Print. Off., 2007. p. 54.

[79] Ibid.

[80] Congressional Record — House, December 17, 2007. p. H16122.
[81] Personal communication, January 15, 2008.
[82] Subcommittee on Nanoscale Science, Engineering, and Technology, Committee on Technology, National Science and Technology Council. 2008. National Nanotechnology Initiative Strategy for Nanotechnology-Related Environmental, Health, and Safety Research. 90 p. [http://www.nano.gov/ NNI_EHS_Research_Strategy.pdf], visited February 15, 2008.
[83] Ibid.
[84] Marburger, John H. III, and James L. Connaughton. Principles for Nanotechnology Environmental, Health, and Safety Oversight, memorandum for the heads of Executive departments and agencies, November 8, 2007. [http://www.ostp.gov/galleries/default-file/ Nano%20EHS%20Principles%20Memo_OSTP-CEQ_FINAL.pdf], visited January 22, 2008.

INDEX

D

E

F

I

J

K

L

M